Palgrave Macmillan Studies in

Series Editor: **Professor Philip Molyneux**

The Palgrave Macmillan Studies in Banking and Financial Institutions are international in orientation and include studies of banking within particular countries or regions, and studies of particular themes such as Corporate Banking, Risk Management, Mergers and Acquisitions, etc. The books' focus is on research and practice, and they include up-to-date and innovative studies on contemporary topics in banking that will have global impact and influence.

Titles include:

Mario Anolli, Elena Beccalli and Tommaso Giordani (*editors*)
RETAIL CREDIT RISK MANAGEMENT

Rym Ayadi and Emrah Arbak
FINANCIAL CENTRES IN EUROPE
Post-Crisis Risks, Challenges and Opportunities

Rym Ayadi and Sami Mouley
MONETARY POLICIES, BANKING SYSTEMS, REGULATORY CONVERGENCE, EFFICIENCY AND GROWTH IN THE MEDITERRANEAN

Caner Bakir
BANK BEHAVIOUR AND RESILIENCE
The Effect of Structures, Institutions and Agents

Alessandro Carretta and Gianluca Mattarocci (*editors*)
ASSET PRICING, REAL ESTATE AND PUBLIC FINANCE OVER THE CRISIS

Dimitris N. Chorafas
BASEL III, THE DEVIL AND GLOBAL BANKING

Dimitris N. Chorafas
HOUSEHOLD FINANCE
Adrift in a Sea of Red Ink

Dimitris N. Chorafas
SOVEREIGN DEBT CRISIS
The New Normal and the Newly Poor

Stefano Cosma and Elisabetta Gualandri (*editors*)
THE ITALIAN BANKING SYSTEM
Impact of the Crisis and Future Perspectives

Joseph Falzon (*editor*)
Bank Performance, Risk and Securitisation

Joseph Falzon (*editor*)
Bank Stability, Sovereign Debt and Derivatives

Juan Fernández de Guevara Radoselovics and José Pastor Monsálvez (*editors*)
CRISIS, RISK AND STABILITY IN FINANCIAL MARKETS

Juan Fernández de Guevara Radoselovics and José Pastor Monsálvez (*editors*)
MODERN BANK BEHAVIOUR

Franco Fiordelisi and Ornella Ricci (*editors*)
BANCASSURANCE IN EUROPE
Past, Present and Future

Josanco Floreani and Maurizio Polato
THE ECONOMICS OF THE GLOBAL STOCK EXCHANGE INDUSTRY

Jill M. Hendrickson
FINANCIAL CRISIS
The United States in the Early Twenty-First Century

Otto Hieronymi and Constantine Stephanou (*editors*)
INTERNATIONAL DEBT
Economic, Financial, Monetary, Political and Regulatory Aspects

Valerio Lemma
THE SHADOW BANKING SYSTEM
Creating Transparency in the Financial Markets

Paola Leone and Gianfranco A. Vento (*editors*)
CREDIT GUARANTEE INSTITUTIONS AND SME FINANCE

Ted Lindblom, Stefan Sjögren and Magnus Willesson (*editors*)
FINANCIAL SYSTEMS, MARKETS AND INSTITUTIONAL CHANGES

Ted Lindblom, Stefan Sjögren and Magnus Willesson (*editors*)
GOVERNANCE, REGULATION AND BANK STABILITY

Gianluca Mattarocci
ANOMALIES IN THE EUROPEAN REITs MARKET
Evidence from Calendar Effects

Bernardo Nicoletti
CLOUD COMPUTING IN FINANCIAL SERVICES

Özlem Olgu
EUROPEAN BANKING
Enlargement, Structural Changes and Recent Developments

Fotios Pasiouras
GREEK BANKING
From the Pre-Euro Reforms to the Financial Crisis and Beyond

Daniela Pîrvu
CORPORATE INCOME TAX HARMONIZATION IN THE EUROPEAN UNION

Ramkishen S. Rajan
EMERGING ASIA
Essays on Crises, Capital Flows, FDI and Exchange Rate

Gabriel Tortella and José Luis García Ruiz
SPANISH MONEY AND BANKING
A History

The full list of titles available is on the website:
www.palgrave.com/finance/sbfi.asp

Palgrave Macmillan Studies in Banking and Financial Institutions
Series Standing Order ISBN 978–1–403–94872–4
(*outside North America only*)

You can receive future titles in this series as they are published by placing a standing order. Please contact your bookseller or, in case of difficulty, write to us at the address below with your name and address, the title of the series and the ISBN quoted above.

Customer Services Department, Macmillan Distribution Ltd, Houndmills, Basingstoke, Hampshire RG21 6XS, England

The Shadow Banking System
Creating Transparency in the Financial Markets

Valerio Lemma
Associate Professor of Law and Economics, Marconi University of Rome, Italy

© Valerio Lemma 2016

Softcover reprint of the hardcover 1st edition 2016 978-1-137-49612-6

All rights reserved. No reproduction, copy or transmission of this publication may be made without written permission.

No portion of this publication may be reproduced, copied or transmitted save with written permission or in accordance with the provisions of the Copyright, Designs and Patents Act 1988, or under the terms of any licence permitting limited copying issued by the Copyright Licensing Agency, Saffron House, 6–10 Kirby Street, London EC1N 8TS.

Any person who does any unauthorized act in relation to this publication may be liable to criminal prosecution and civil claims for damages.

The author has asserted his right to be identified as the author of this work in accordance with the Copyright, Designs and Patents Act 1988.

First published 2016 by
PALGRAVE MACMILLAN

Palgrave Macmillan in the UK is an imprint of Macmillan Publishers Limited, registered in England, company number 785998, of Houndmills, Basingstoke, Hampshire RG21 6XS.

Palgrave Macmillan in the US is a division of St Martin's Press LLC, 175 Fifth Avenue, New York, NY 10010.

Palgrave Macmillan is the global academic imprint of the above companies and has companies and representatives throughout the world.

Palgrave® and Macmillan® are registered trademarks in the United States, the United Kingdom, Europe and other countries.

ISBN 978-1-349-57812-2   ISBN 978-1-137-49613-3 (eBook)
DOI 10.1057/9781137496133

This book is printed on paper suitable for recycling and made from fully managed and sustained forest sources. Logging, pulping and manufacturing processes are expected to conform to the environmental regulations of the country of origin.

A catalogue record for this book is available from the British Library.

A catalog record for this book is available from the Library of Congress.

Contents

List of Tables   viii

Introduction   1

1 General Observations   13
  1.1 The identification of the phenomenon   13
  1.2 The traditional definitions of the shadow banking system: the guidelines of the FSB and the statement of the G20   17
  1.3 The directions of certain central banks   21
  1.4 The routes of European institutions   23
  1.5 The path of emerging countries   26
  1.6 The interpretations of the phenomenon   26
  1.7 The different outcomes of the monetary and supervisory perspectives   30
  1.8 The boundaries of the shadow banking system   31
    1.8.1 Money laundering, tax evasion, and other forms of "black market"   31
    1.8.2 Prohibited shadow operations   35

2 The Shadow Banking System as an Alternative Source of Liquidity   37
  2.1 The economic determinants of the shadow banking system   37
  2.2 Information asymmetries   42
  2.3 Opacity, pro-cyclicality, and system instability   44
  2.4 Methods for classification of the phenomenon   48
  2.5 Is this economic freedom or escape from regulation?   53
  2.6 The global nature and the riskiness of the phenomenon   55
  2.7 New freedoms and their problematic nature   58

3 Shadow Banking Entities   62
  3.1 Special purpose vehicles   62
  3.2 Shadow banks   67
  3.3 Other shadow banking entities   70

	3.4	Shadow funds	73
	3.5	The particular role of money market funds	75
4	**Shadow Business of Banks, Insurance Companies, and Pension Funds**		78
	4.1	Multiphasic shadow credit intermediation process and the roles of traditional operators	78
	4.2	The role of banks in the shadow banking system	81
	4.3	The dysfunctions of internal controls and weaknesses of other safeguards	85
	4.4	The action of insurance companies	87
	4.5	The involvement of pension funds	92
5	**Shadow Banking Operations**		95
	5.1	Shadow credit intermediation process	95
	5.2	Undetermined (contents) and unconfined (boundaries) of shadow banking operations	98
	5.3	The operations of the credit transformation	101
	5.4	New securitization techniques	102
	5.5	Liquidity, maturity transformation, and financial leverage	105
	5.6	Ratings (in the shadows)	107
6	**Non-Standard Operations in the Shadow Banking System**		113
	6.1	The use of securities lending and borrowing, and repurchase agreements	113
	6.2	The offering of the asset-backed commercial paper, asset-backed securities and collateralized debt obligations	117
	6.3	Peculiarities of derivatives	120
	6.4	The shadow banking operations of sovereign states	124
	6.5	The shadow banking operations of credit institutions	127
7	**Shadow Banking Risks and Key Vulnerabilities**		130
	7.1	Areas of risk in the shadow banking system	130
	7.2	Risk factors	133
	7.3	Operators and internal policies of risk management	136
	7.4	Operational freedoms inside the shadow banking system	139
	7.5	The risks of organizations	140
	7.6	The risks of governance	142

7.7	The impact of European regulation on the shadow banking risks	144
7.8	The risks of the entities "too big to fail"	145
7.9	The exogenous risks	147
7.10	The particular implication of monetary policies	151

8 The Shadow Banking System and the Need for Supervision — 154

8.1	Checks and balances in the shadow banking system	154
8.2	Economic determinants of the supervisory system on shadow banking	156
8.3	The shadow banking system in the European internal market	158
8.4	The role of European institutions	160
8.5	New supervision on the shadow banking system in the European Union: the action of the European Commission	161
8.6	The action of the European System of Financial Supervision	163
8.7	The responsibilities of the European Supervising Authorities	165
8.8	The tasks of the European Banking Union	167
8.9	The impact of the new targeted longer-term refinancing operations of the European Central Bank	171
8.10	Evolutionary trends of European supervision (following Directive 2014/65/EU)	172
8.11	The role of global regulators: the World Bank and the International Monetary Fund	175
8.12	Limited effects of the supervision designed by the Financial Stability Board	178

Conclusions — 182

Notes — 185

Index — 235

List of Tables

1.1	The phenomenon in the capital market	17
1.2	Definitions	18
1.3	Boundaries	33
2.1	Determinants of the shadow banking system	39
2.2	Economic perspectives	54
3.1	Features of shadow banking entities	63
4.1	The roles in the shadow business	81
5.1	Shadow credit intermediation process and regulated activities	96
7.1	Areas of risk	132
7.2	Risk assessment	134
8.1	Supervising network	158

Introduction

A clear definition of the phenomenon under observation, *shadow banking*, is not appropriate at the beginning of an analysis like the following. After all, we are dealing not with the essence of the banking business, but with the conditions and the effects of capital circulation. As the following pages of this book will show, the external course of the shadow banking system is multidimensional, and hence we will be able to understand its features and risks by analyzing the concrete evidence of experience.

The object of this study is, therefore, the "economic space" that lies between the financial market and the illegal circulation of capital. This is a residual field, with limits marked by the choices of liberalization in the credit intermediation activities carried out by the legal systems of developed countries.

Our interest in the shadow banking system arises from its economic relevance, because—according to the quantitative analysis published by the Financial Stability Board—it influences the industries of the G20 countries (and the whole euro area), being connected with 80 per cent of global gross domestic products (GDPs) and 90 per cent of outstanding financial assets.[1]

It is useful to take into account the first economic analyses, which were aimed at examining the role of the shadow banking system (in relation to the financing of the real economy and, at the same time, to the proliferation of micro-and macro-systemic risks). These analyses have focused on credit transactions with direct execution on the financial market and, therefore, without the involvement of a bank.[2] As a result, it is clear that these transactions can facilitate the "movement of capital" according to different parameters to those established by prudential supervision.

Shadow operations seek out subjects with mutual interests who, for diverse reasons, choose to operate outside the banking rules (without breaking them). These operations are based upon contracts that are able to sustain a lawful "sequential trading system" (different from the traditional banking business). Therefore, the book will not focus on the operations carried out in breach of the market regulation, as they are active in the "black market" (and not in the shadow banking system).[3]

In other words, the shadow banking system will be explored as it is used in cases that allow a "subject in need of credit" to obtain (long-term) financing, and another "subject in need of investment" to achieve financial instruments (short term). It will be considered that this is related to the underwriting of contracts, their securitization and re-securitization. These operations involve certain *special purpose vehicles* (SPVs), which are equipped by the executives who arrange the transactions (in order to connect the initial loans with the final *collateralized debt obligations*).[4]

It is opportune to anticipate that the financial instruments issued by several subjects will be stored together in the assets of these SPVs, by bringing to the table a "mixture of assets" that are repeated at every stage of the *shadow credit intermediation process*. Thus, a loan and asset-backed security (ABS) and asset-backed commercial paper program (ABCP) warehousing situation will be determined within the assets of an SPV active in the shadow banking system. These assets are, ultimately, composed of an underlying set of different debt situations, joined together by the common referability to the same "shadow banking operation."[5]

The financial instruments issued by the first vehicle (so-called ABCP) are subscribed to by a second vehicle, which will issue other debt securities (so-called ABS). The operation then repeats itself again until a last vehicle issues the final financial instruments (the so-called collateralized debt organization or CDO).

Obviously, at each phase, an operation of ABS warehousing is also carried out, hence the assets of a vehicle will be composed by different financial instruments (with regard both to the person who issued them and the economic content). It should, therefore, be considered that each phase follows another, and at any stage the abovementioned operations are repeated.

It goes without saying that the shadow credit intermediation process is the "chosen procedure," which gives life to a special form of *market-based financing*. In fact, this process determines a complex situation that can repeat an indefinite number of operations (in the above sequence, that is from the provision of loans, to the issuing of securities, to the placement of CDOs).

In each of these phases, several effects may be produced: leverage (since a vehicle is free to determine the level of its financial leverage), maturity transformation (without correlating the expiry of the securities to the duration of the loans), and securities recalculation (proceeding with the issuance of more classes of securities, with different capital guarantees, so-called credit transformation).[6]

These are the reasons why I will refer to this alternative banking channel as "shadow," on the understanding that, in the case of the financial markets, the collection (or possession or transferring) of capital shall not imply that the public authorities are required to play a supervisory role.

It is also useful to take into account the derivatives put in place by the operators of the shadow banking system. This practice enables both the effects and the synthetic circulation of the loans provided to the real economy.[7]

Therefore, special attention should be paid to the contracts that allow the exchange of cash flows (i.e. swaps), the predetermination of certain execution conditions of the transactions (i.e. futures and options), and the other rights that you can get through the conclusion of a derivative (even with supervised intermediaries). These derivative contracts, in fact, do not exhaust their impact within the economies of the entities that are part of them, but extend their influence to the industry under investigation (in terms of fluidity and riskiness).[8]

This suggests the systemic importance of such derivatives, which—as it will be seen—create the need to provide appropriate safeguards (of the fulfillments) and transparency (of the information), in order to avoid systemic effects (in case of default of an operator).

From another perspective, it is necessary to have regard for the analysis of the technical profiles of the shadow banking system, in order to verify that we are not in the presence of a mere secondary phenomenon (in terms of socio-economic interest and systemic effects), nor a sophistication of the mechanism of securitization (which is also at the basis of the operations that give content to this case).[9] It seems, rather, that the empirical analyses have detected a new system (as an alternative to the banking one), which goes beyond the traditional mechanisms of capital circulation and payment systems (regulated by the EU authorities).

It is, therefore, necessary to verify whether, at a legal and economic level, the shadow banking system is compatible with the approach of the social democratic order of the most developed countries. Hence, there is the need to clarify whether access to such a system is pre-ordained to the reduction of transaction costs (generated by banks) and the achievement of additional benefits (consistent with the rights of other citizens). Only in this way could it be ruled out that, in these shadows, opportunistic behavior can make a proliferation of unacceptable dangers for the global financial system possible (i.e. by diverting resources from the real economy to infect the financial markets and to put in crisis the traditional banking channels).[10]

Accordingly, the economic determinants of the shadow banking system must be determined and, therefore, their compliance with the legal criteria that informs the subject of financial supervision. This identification is aimed at understanding if the recourse to these "collateral circuits"—to improve the financing of the real economy—is compatible with the ethical values of the referential legal system.[11]

There is no doubt, in fact, that the alternative choice (between shadow and traditional banking) should not (in any way) allow an irregular circulation of money and, therefore, a circumventing of the safeguards that any democratic system poses to ensure a balanced protection of its citizens' rights (economic and financial).

In this context, a legal and economic assessment of the shadow banking system—and of the need to provide regulation and control interventions—must, on the one hand, have regard to the renewed structure of financial supervision, and, on the other hand, be able to weigh up its (positive or negative) externalities. This assessment will be aimed at identifying the methodologies that are able to guide public intervention towards the optimality of the results (evaluated in reference to the "social welfare criterion").[12]

That said, it is necessary to identify the "elements" of this financial architecture that are located in a regulatory space between the boundaries of the controlled and the prohibited transactions. Consequently, the processes of globalization of society and financialization of the economy must be taken into account. Then, the democratic legitimacy of this system needs to be addressed, in order to acknowledge the presence of entities withdrawn from the special supervision (and, therefore, excluded from the application of the safeguards that occur in this case).

The result of this verification will depend on the assessment of the socio-economic function of the shadow banking system. In fact, in this book, I will consider the possibility that this is a system in which, on the one hand, you can negotiate an intermediation margin lower than that commonly practiced by commercial banks (because of the correlation to operating costs lower than those of the latter) and, on the other hand, a lender can take higher risks and charge greater interest (unsustainable by intermediaries subject to the regime of "capital adequacy" provided by the Basel Agreements).

Following the recent financial crisis, the international institutions and monetary authorities show growing interest for the mechanisms of wealth circulation that, without the intervention of banking intermediaries, achieve a form of "credit intermediation involving entities and activities outside the regular banking system."[13] This interest was related not only to the quantitative importance of these varied forms of market-based financing, but also to the externalities that they produce within the global financial system (in terms of proliferation of systemic risks).

This is a perspective that focuses on the subjects not included in the scope of government supervision, and on the operations aimed at achieving a new way for capital circulation (beyond market rules and supervision). And it is in this perspective that the shadow banking system awakens the interest of the international community.

Therefore, it is necessary to start our analysis from the conclusions of the G20 Seoul Summit of November 10–11, 2010. In this summit, the representatives of the gathered countries made the commitment to develop macroeconomic policies so as to ensure ongoing recovery and sustainable growth, in order to enhance the stability of financial markets (the so-called "Seoul Action Plan"). This commitment explains why the first significant "common action guidelines" were established to overcome the *imbalances* that undermine the possibility of an economic recovery. Moreover, these guidelines were founded on the economic determinants at the basis of the international regulatory framework for banks (created by the Basel Committee on Banking Supervision), in order to strengthen supervisory oversight of the global financial system by using new forms of control on the shadow banking system.

It is clear that these directions go beyond the previous regulatory approach, in which the stability of the market was pursued through the safety of the supervised institutions. A new methodology, designed to prevent financial imbalances caused by the asymmetry between the conditions in which banks operate and those in which other forms of credit intermediation, is implemented (i.e. affirmation of the shadow banking system).

Additionally, the option for this new method corresponds to the task of filling a regulatory gap that for too long has jeopardized the smooth functioning of the market. This option was supported by the political decision "to raise standards, and ensure that our national authorities implement global standards developed to date, consistently, in a way that ensures a level playing field, a race to the top and avoids fragmentation of markets, protectionism and regulatory arbitrage."[14]

The conclusions reached in the following G20 Summit in Cannes on November 3–4, 2011 are in line with the abovementioned directions. There is a confirmation of the centrality of the regulatory issues raised during the crisis. And, in agreeing the "Seoul Action Plan," the G20 leaders specify the need to proceed with the development of a new system of clearing and trading obligations for over-the-counter (OTC) derivatives. The setting of standards and principles for sounder compensation practices correlate with the satisfaction of this need.

This is a clear commitment to adopt a single set of high-quality global accounting standards, able to realize a comprehensive framework, in order to address the risks posed by systemically-important financial institutions (SIFIs).[15] It is possible to identify the beginning of a regulatory process for the implementation of strengthened rules and oversight on shadow banking—through the control of the business relationships that the supervised intermediaries have with the shadow banking entities. An attempt to exploit the effectiveness of traditional supervision mechanisms to handle the complexity of the global financial system can also be ascertained.

It goes without saying that the leadership of the Financial Stability Board (FSB) is integral to such an innovative process. As will be shown, the role of this board goes far beyond the task of collaborating with other international standard-setting bodies to define the rules necessary to strengthen the supervision of the shadow banking system. And, sometimes, it also goes beyond the directions of the G20.

The new method of intervention is, therefore, based on inter-institutional collaboration and technical co-ordination (since the adoption of "The Los Cabos Growth and Jobs Action Plan"). This also implies that the FSB has a responsibility to monitor the program of reforms proposed by the G20 in its summits, as it has to report to the FSB's "Coordination Framework for Implementation Monitoring" (CFIM).[16]

However, the limitations of such a structure are evident: the powers of the FSB come up against the "sovereign barriers" inherited from the twentieth century. The commitment of the FSB cannot, in any way, go beyond the reporting of the independent choices made by national regulatory authorities appointed for the matters covered by the aforementioned G20 program. Moreover, there is no trace in the contemporary world of a global legal order, nor of a sectorial regulatory regime that can confer more powers to the FSB. Hence, the institutional framework is inadequate to manage the network of cross-border interdependencies generated by the shadow banking system and, therefore, the international organizations find a practical difficulty in promoting a common regulation to overcome the opacity of the system and preventing market failures.

The G20 member countries show significant interest in this context. These countries have embarked on an individual path of regulation to the phenomenon. Noted in particular is the fact that, at the G20 St. Petersburg Summit on September 5–6, 2013, the United States explicitly stated their intention to reduce the risks emanating from the shadows and, therefore, to increase the prudential standards (for banks or designated non-banks), in order to address vulnerabilities in short-term wholesale funding markets. This statement is in line with the conclusions agreed by all the G20 leaders, according to which the supervision must proceed beyond the adoption of the rules, designed in order to ensure a more resilient banking industry, by promoting the improvement of transparency and fairness, and by filling regulatory gaps that allow the proliferation of the risks in the shadow banking system.

The will to deal with the various problems raised by the crisis has been made explicit; starting with the extension of the public intervention. Recently we have seen instances of a direct intervention on the shadow banking system, and for the implementation of a monitoring system (on the shadows' subjects or activities). Today, a proactive orientation has begun to find affirmation, which is aimed at making the law and its principles penetrate in the space occupied by the phenomenon under observation.

Therefore, this study aims to achieve a better understanding of how the problems posed by the shadow banking system (and their manifestation on a global scale) will be tackled.

The progress made by the G20 is remarkable. Today the G20 is set up as a body that is able to elaborate policy recommendations on financial matters, even if it doesn't have the power to promote the strengthening of the oversight and regulation systems. The same holds true for the role of the FSB, a group that is certainly far from being an authority responsible for the regulation and control of affairs (financial and monetary).

This progress, from the first observations of the phenomenon to date, is in line with the international community's intention to regulate the presence of an alternative conduit able to increase the volume of credit made available for the real economy. Hence, I will analyze and find out which rules are useful to increase the transparency of market relations and guarantee the efficient management of systemic risk arising from the shadows. This goal will lead to new paths that can be followed to ensure the smooth operation of the shadow banking system.

In light of this introduction, it seems preferable to conduct the investigation in a perspective that takes into account the financial supervision and the interventions by the monetary authorities. It is clear, in fact, that both shall not apply directly to the shadow banking system.

Therefore, it seems useful to move from the assumption that the phenomenon under observation is outlined in consequence of the tendency of the financial operators to reduce every type of cost, including those incurred to comply with supervisory expenses. To this trend corresponds, consequently, the attempt to operate "outside of the rules," without it being considered by the authorities as a violation of the referential financial legal order. Hence, from this point of view, the shadow credit intermediation process is the final outcome of the search for alternative ways (for the circulation of capital). And this is true both in terms of risk and performance.

In addition, I will evaluate the impact of other economic determinants of the phenomenon that, as shall be shown, are related to the flexibility of the organizational structure of the shadow entities (on which less transitive and submerged costs bear down). Therefore, I will investigate under which conditions the shadow banking system is a successful model for the temporary transfer of wealth outside of the banking channels.

Dwelling on the examination of the phenomenon from an investor's point of view, it will be observed that the relevant financial instruments (issued to support the process of intermediation) offer advantageous conditions of risk remuneration (given the aforementioned lower incidence of costs). Furthermore, they provide the possibility (for investors) to buy securities riskier than those offered by the banking sector (and, therefore, with a higher return rate). These financial instruments are suitable to meet the needs of operators looking for opportunistic chances of individual profit.

There is no doubt that such a system is exposed to the risk that its operating conditions do not reach an efficient level of safety, such as to maximize social welfare or at least to avoid economic losses caused by imbalances in the relevant industry (so-called deadweight loss).

I will, therefore, keep in mind that the option for high-risk shadow operations occurs in the absence of safeguards aimed at avoiding information asymmetries and other conditions of *opaqueness* of the market. And this casts doubt on the fact that, in the shadow banking system: (i) the meeting between supply and demand is the result of a "cognitive activity" aimed at the acquisition of in-depth informational data, and (ii) the investment choices are, therefore, the result of rational decision-making processes.

Another issue concerns the compatibility—or not—of such a risky system with the modern democratic legal order, and then with the safeguards introduced for the protection of the "common good" and the purpose of "social utility." However, during the investigation, this issue will not be reduced to the problematic definition of the freedom at the basis of the intermediation activity, always contended between entrepreneurial autonomy and public control. Instead I will proceed to the clarification of the doubt that the access to this parallel system is (only) an exercising of operators' freedom and not (also) the poor behavior of people looking for regulatory arbitrages.

Thus, deeper investigations—that go beyond the verification of the specific economic reasons related to the convenience of the transaction placed in the shadow banking system—are necessary. In fact, the book is concerned with both the analysis of the subjects that operate in this system and the study of the operations that allow the circulation of wealth from those in surplus to those in deficit.

The multiphasic nature of the shadow banking system determines the involvement of a number of operators. Many, in fact, are the skills required to achieve a form of credit intermediation that takes place according to a sequential process whereby, once securitized, the initial funding phases (in which varied financial tools are issued and stocked) alternate until the placement of the CDOs in the financial market.

It is certain that the phenomenon under observation has variable borders, including new types of operators if an evolution of the sequence of operations is registered. Therefore, the examination of the shadow banking entities will take into account all those who participate in the shadow credit intermediation process.

In this context, I connect the presence of SPVs to the type of operations used to achieve the credit intermediation. In fact, these vehicles result in being ancillary to the transformations that allow the offering of financial instruments in the configuration best suited to the demand made by the investors. Conversely, it will be difficult to identify a true "shadow bank" in the system under observation, because there is not an individual author of credit intermediation (made outside the spheres of public supervision, and in the absence of the direct relationships with the monetary authorities).

With particular reference to banks and other investment firms, it appears clear that their participation in the shadow banking operations is fully justified on an economical basis. Indeed, these companies can provide (banking or financial) services in favor of the abovementioned shadow banking entities; these services appear to be fully in line with the objective of increasing the volume of activity, and in turn with the goal of improving the performance of the business. The same has to be said for the specific operations of lending (in any form), where their validity must be assessed in relation to the individual conditions of the loan and the creditworthiness of the recipients.

However, an analysis will be provided of the "originate-to-distribute" banking business model. We have to understand the links between the shadow banking system and the credit institutions that, themselves or through related entities, were involved in an original lending agreement that created the obligations of the debtor giving rise to exposures to securitized (according to the provisions of EU Regulation no. 575/2013, art. 4, para. 1, point 13). The same is true for the banks that purchase third-party exposures for their own accounts and then securitize them. In both cases, these banks do not hold the credits in their assets (as happens in the "originate-to-hold" model).

I will then focus on the procedure used to place the original exposures in the shadow banking system, because these credits are the "raw material" for the production cycle of CDOs. In such a case, the quality of the information collected by the banks (on the creditworthiness) will have to guarantee the reliability of the aforementioned credits (and, therefore, the value of what underlies the financial instruments issued at the end of the shadow credit intermediation process). It will be the cash flow generated by these credits that will reward all the operators and ensure an interest from those who subscribed to the CDOs.

In light of the foregoing, the importance of the analysis concerning the shadow banking operations is understandable. The focus will then turn to the operations that transform the loans (granted to the real economy) in the financial instruments (offered to the capital market).

As a consequence, whether or not the tendency of the shadow banking system to evade the rules affects the influences of the phenomenon under observation must be clarified. This assessment will guide the analysis towards the verification of the sustainability of this unregulated financial reality. The *fear* that we are in the presence of an inefficient context (both in terms of result and of resource allocation) cannot be ignored. Hence, the book's intention is to assess whether the new directions of the international authorities can promote new appropriate solutions in order to improve the quality of the circulation of capital (outside of the banking sector).

Innovation and globalization are the drivers of the shadow banking system's success. Only as a result of their development, in fact, can the technical conditions required for the execution of the shadow banking operations be achieved. I refer, in particular, to the freedom of cross-border movements of capital, to the opportunities offered by the application of financial engineering, and to the possibility of choosing the venues of financial instruments.

In turn, the development of a "global financial system" allows new players to revolutionize the intermediation processes, creating a new financial capitalism.[17] I refer also to the process of financialization of the economy that led to the spread of overly structured and complex financial instruments. And both these phenomena go along cross-border networks. It will be necessary, therefore, to give careful consideration to the outcome of the deregulation process, as it followed the freedom (in the application of the financial innovations) that has marked the end of the twentieth century and the beginning of this new millennium.

The result is an anomalous trend of the financial industry (that we have been calling "financial crisis" since 2007). Only at first did the market trend go along with the development of the real economy, increasing the velocity of the circulation of money and improving other factors that determine the volume of the credit supply. Later, in fact, the behaviors have become increasingly careless, allowing operations undocked from any prudential parameter. Hence, a specific interest for the role played by intermediaries and other firms in a financialized economy is developed, because their performance influences the provision of reliable services and the use of instruments able to ensure the safe management of wealth.

The current high level of financialization (of the economy of the G20 countries) confirms the importance of these aspects in the analysis of the shadow banking system. Thus, the effects of the growing volume of financial assets traded in the capital market need to be understood.

As a consequence of the above, this analysis will examine whether the affirmation of the shadow banking system represents one of the effects of a regulatory policy that, in pursuing the maximum welfare, relies on free markets, private property, and minimum interference by the public sector. In the affirmative, the freedom of the shadow banking system will appear as a political decision, a regulatory option, a consequence of a growing dominance of market-based financial systems over bank-based systems.

It must be taken into account that, in doctrine, the theses that define financialization as the increasing dominance of the finance industry are not rare.[18] However, we must also consider that the most developed countries show greater interest in the monitoring and supervising of any banking system, and then that their national authorities shall improve certain safeguarding factors for the alternative mechanisms of capital circulation (aimed at the prevention of systemic risks).

Therefore, the book will examine closely the role of the regulators, as they influence preferences regarding the allocation of investments. I will verify whether or not the investment of private savings into this system is linked to the "choice between saving in the form of cash, bank deposits or stocks, or perhaps a single-family house, [which] depends on what one thinks of the risks and returns associated with these different forms of saving."[19] At the same time, I will consider that—as it has been demonstrated by Fama, Hansen, and Shiller—the "asset prices" provide essential information for this rational process of investment.[20]

However, it will also be taken into account that there are other variables concerning the shadow industry's levels of transparency, security, and stability. In the shadows, in fact, these variables shall determine the market prices (of financial assets) and their link to the underlying values. Indeed, there is an awareness that the absence of rules can determine the mispricing of assets and, consequently, contribute to the triggering of crises (as it was in the US subprime mortgage crisis).

Under these assumptions, I will consider that the regulation of the shadow banking system shall face certain problems of the global financial system, since I agree with the scholars that state: "we do not yet have complete and generally accepted explanations for how financial markets function" and "the question of whether asset prices are predictable is as central as it is old."[21] Only by solving these problems can the regulators hope to avoid general prohibitions of the open-market operations (of loan warehousing and securities issuance), because this prohibition could lead to the *obsolescence* of the shadow credit intermediation process, but not to a new model for a safer alternative circulation of capital.

On the contrary, in this book, I prefer a solution able to ensure the efficiency in managing high risks, by supervising information transparency, risk awareness, and fair compensation for risk taking. These actions will ensure the survival of the phenomenon under observation, and then the results achieved as a consequence of the development of global markets' networks and financial innovation.

1
General Observations

Under a regulatory approach, this chapter will address the questions about the definition of the shadow banking system and its basic elements.

First the boundaries of this system will be defined; then I focus my attention on the definitions provided by international bodies: the Financial Stability Board (FSB), the International Money Fund (IMF), the World Bank—and national authorities: the Federal Reserve System (FED), the Bank of England, the Bank of India, the Bank of China, the Bank of Italy—from both a monetary and supervisory point of view.

The chapter will go on to explain the differences between the shadow banking operations and the illegal practices (i.e. money laundering, tax evasion, etc.) in order to understand the benefits of this market-based form of credit intermediation.

1.1 The identification of the phenomenon

In countries with a high rate of financialization, the legal systems do not provide for a "fundamental rule" that defines the concept of the shadow banking system. The shadow banking system is, therefore, not considered as a whole.[1] Specifically, it will be observed that the European and national regulations are considering this phenomenon with regard to its empirical manifestation, arising from the choice of private individuals to let their capital circulate outside of a regulated financial market or without the involvement of a credit institution (as the banking system is defined by art. 3 of Directive 2013/36/EU).

Consequently, it is necessary to identify the elements that lay at the foundation of this choice, but without researching—within the shadow banking system—an "exchange center" towards which the open market operations converge. Here I move away from the assumption that the present situation does not reflect the known concepts of "regulated market," "multilateral trading facilities" (MTF), and "organized trading facility" (OTF), used by Directive 2014/65/EU, to identify "trading venues" that must comply with the requirements of European law (and, therefore, the provisions of the European financial supervision system).[2]

Moreover, the financial transactions that are not referable to a single country and, therefore, to a specific system of supervision will be taken into account. These transactions are not included even in the perimeter of the new surveillance mechanisms that go beyond national boundaries (in the forms, first, of the Sevif and, then, of the European Banking Union). Neither of these mechanisms seems sufficient to encompass the phenomenon in its entirety. It is, therefore, penalizing the absence of a "global supervision" that extends its scope to the global financial network that crosses the planet (and, therefore, exceeds the obvious territorial limitations of the sovereign authorities).

Therefore, the financial markets demonstrate the existence of a network of legal relations and the absence of an authoritative system of regulation and control (dedicated to them). These conditions are sufficient to identify the core foundation of the shadow banking system. This is why specific elements are identified to support a legal reflection about future actions—useful or necessary—in order to ensure that the freedom of the shadow system should not damage the public welfare.

It appears, however, necessary to keep in mind that regulatory actions could achieve inefficient results if the cost of the supervision is more onerous of its benefits. In other cases, the shadow operations could avoid general interests and, then, certain specific prohibitions are required (even if they are not efficient).

That said, it is useful to bear in mind that, in economic terms, there is a functional definition of the shadow banking system, which is able to allow a measurement of its quantitative importance. And, indeed, any operation executed out of any supervised trading venues can be accounted to the shadow banking system. This is, therefore, part of the financial global system, which, to date, allows the movement of capital across the world, through the channels of an integrated global network, dynamic and innovative, formed by interactive components consisting of: intermediaries, securities (products/instruments), markets, derivatives, regulation and supervision, payment, clearing, and settlement systems.[3]

In addition, the relationship between the economic importance of the phenomenon under observation and its multilateral nature is especially relevant; because it is based on a systematic meeting of interests (which are referable to several subjects, some of them in surplus and others in deficit of capital). This is, in fact, a characteristic trait, which excludes from the scope of our investigation operations that are carried out on a bilateral basis (in reciprocal and symmetrical modes).

Both of these elements are implicitly implied by the term itself—"shadow system"—where the assumption is clear that the systematic character of the operations shall be realized in terms of opacity. The word "banking," then, suggests the selection of only those operations that allow the collection of resources for the start of any credit business. That said, it is necessary to dwell on the meaning attributed to the banking qualification for such a shadow system.

The use of the term banking, in fact, reflects the specificity of the interests that are part of the system under consideration. This also highlights the nature of the goals of price stability, savings protection, and credit control, which conform to the legal requirements of countries with advanced economies. Moreover, as a result of the recent financial crisis, these goals influence the evolutionary process of European law, which, as we shall see, is now characterized by new common forms of supervision (and prudential vigilance) on the banking and financial sector—that is, the European System of Financial Supervision (ESFS) and the European Banking Union (EBU).[4]

In light of the foregoing, we can understand that the banking qualification of this intermediation process of capital can be considered both as a constitutive element of the system and a peculiar configuration of its operations—which lets subjects with surplus resources enter into a relationship with others that, to meet their needs, are willing to remunerate the temporary transfer). This applies, in particular, to the various operations that are able to overcome the barriers to the meeting—in traditional ways—of demand and supply of capital (referring to the asynchrony of the deadlines, the asymmetry of the risk profiles, or other impediments, including ones of geopolitical nature.

In this context, a complete definition of the shadow banking system must take into account the subjects not included in the scope of government supervision and the operations aimed at achieving a (regular, synthetic, or derivative) circulation of capital outside of any regulated market, without the involvement of supervised intermediaries.

In brief, it can be said that we are not considering the operations based on organic relations (and, therefore, those made within a single subjective entity) or intra-group (or rather, within the same socio-economic unit). The same can be said with regard to other activities that do not comply with an intermediary logic, but pursue speculative goals. In particular, we cannot include in the shadow banking system any operation or action if it takes resources from the economic system, instead of intermediating them (and then does not multiply them through rapid circulation). There is no doubt, in fact, that these operations and activities—while they should not be regarded as degenerative—are predatory towards the goals of social utility that lie at the root of every balanced and democratic system.

The aforementioned considerations lead to the reconnection of shadow banking to the effects of financial liberalization, which do not prevent economic operators (i.e. companies and investors) from negotiating with subjects that intermediate capital without applying the safeguards provided by supervision.

Addressing the analysis of the phenomenon from this perspective, it appears possible to highlight the importance of the operations of securitization and the wholesale funding, as well as other methods aimed at increasing the raising of capital:

- asset-backed commercial paper (ABCP) conduits;
- structured investment vehicles (SIVs);
- credit hedge funds;
- money market mutual funds;
- securities lenders;
- limited-purpose finance companies (LPFCs); and
- government-sponsored enterprises (GSEs).

The same has to be said for the relevance of the:

- government-sponsored shadow banking subsystem;
- internal shadow banking subsystem; and
- external shadow banking subsystem.

Therefore, this is a complex phenomenon in which—as shall be shown in the following chapters—there are several structures and subsystems. These denote original profiles, which influence the individuals' mutual business relations. Nonetheless, this is a system that is composed by multiple exchange relationships—bilateral and multilateral—all interconnected (and, therefore, included in the definition of shadow banking system).

The consequence is an independence of the shadow banking system from the traditional economic structure. This is exactly the foundation

Table 1.1 The phenomenon in the capital market

Capital market
Market-based financing industry
Shadow banking system

for a rising trend and, then, for a new system of private relations (where the abovementioned structures do exist only if the individual chooses to link the credit intermediation process with them). Hence the power of the structure of the single transaction over the whole system, since the latter can be considered as the result of free market relations.

This suggests that shadow banking can be also a *discontinuous set* of heterogeneous processes, employed by several individuals; resulting in an informal coordination of financial structures (see Table 1.1).

There is, in other words, an accumulation of "entities and operations," which tend to constitute a system. This corresponds to the intensity of the relationship between the subjects engaged in it, determining a sort of *cohesion* that may cause significant systemic effects.

This leads to the conclusion that: in the shadow banking system, *prima facie*, it is possible to relate the opacity of its qualification (i.e. the term shadow) to the collateral position that it occupies (in relation to the regulated markets subject to government supervision). It is, at least in part, determined as a result of a specific choice made by the operators (and, that is, by the exclusion of certain operations from the burden of the legal rules and safeguards of public supervision). However, it can be also a consequence of certain dysfunctions and delays in the regulatory process, with obvious implications—in terms of accountability—for the authorities who fail to renew their (economically obsolete) practices and standards.

1.2 The traditional definitions of the shadow banking system: the guidelines of the FSB and the statement of the G20

Priority must be given to the economic analyses that, in monitoring the shadow banking system, have identified its essence in the "credit intermediation involving entities and activities outside the regular banking system."[5] These analyses allow the interpreter to identify a particular form of market-based financing, which accounts, in quantitative terms, for the growing economic importance of the transactions that develop through it and, in terms of quality, for its *foreignness* to the paradigms of prudential supervision and capital adequacy.

The FSB formulated the abovementioned definition in residual terms—in comparison to the boundaries of the traditional banking system.[6] This means that the FSB did not highlight the constitutive elements of the phenomenon. And indeed, it does not address the benefits brought—to the real economy—by this alternative credit intermediation (i.e. placed outside of the banking channels), nor does it indicate explicitly that this form of intermediation can also have risk–reward parameters that are not compatible with the system of prudential supervision (see Table 1.2).

Hence, the analysis starts from the lack of clarification about the possibility that—within the shadow banking system—it will generate (and will proliferate) systemic risks arising from: (i) the transformation of the maturity, (ii) the exploitation of leverage, (iii) the frequency of the relationship between the industry operators, and (iv) the inefficiency of the risk transfer mechanisms.

Table 1.2 Definitions

FSB	• 2014 • Credit intermediation involving entities and activities (fully or partly) outside the regular banking system, or non-bank credit intermediation in short.
IMF	• 2013 • Many financial institutions that act like banks are not supervised like banks. (Kodres)
ECB	• 2012 • Activities related to credit intermediation, liquidity, and maturity transformation taking place outside the regulated banking system. (Constâncio)
Federal Reserve	• 2013 • Shadow banking activities consist of credit, maturity, and liquidity transformation that take place without direct and explicit access to public sources of liquidity or credit backstops. (Pozsar – Adrian – Ashcraft – Boesky)
Bank of England	• 2010 • Instruments, structures, firms, or markets which, alone or in combination, replicate, to a greater or lesser degree, the core features of commercial banks: monetary or liquidity services, maturity mismatch, and leverage. (Tucker)
Bank of Italy	• 2013 • A "securitized banking" business model, in which loans were distributed to entities that came to be known as "shadow" banks. (Meeks – Nelson – Alessandri)
Deutsche Bundesbank	• 2014 • The shadow banking system comprises all entities and activities that are involved in credit intermediation outside the regular commercial banking system.

In 2014, the FSB presented new results—in its fourth annual monitoring exercise using end-2013 data—showing the danger that excessive competition may turn the terms of credit intermediation against certain operators. Moreover, the FSB highlighted that its new exercise will influence future monitoring reports, understanding that the "narrowing down approach" shall leverage on the new information sharing standards introduced by EU rules (on markets, financial firms, and credit institutions that operates with shadow banking entities).

This exercise may be criticized for the apparent indifference to the new data requirements introduced, at European level, by Emir Regulation (no. 648 of 2012). However, there is an awareness that the realism of the FSB's Workstream on Other Shadow Banking Entities (WS3) is much preferable to the confused "first-run data accounting" of the new "trade-repositories" (which centrally collect and maintain the records of derivatives, under the direct supervision of ESMA).[7]

Consequently, it is easy to understand the need for a more accurate monitoring intervention (by the competent authorities) that is able to prevent the occurrence of systemic crises, especially because of the supranational nature (or rather, cross border) of these *shadows*[8] and the implications due to the absence of a unique regulatory framework (uniformly applicable to the operations and the parties that give content to the system in question).[9]

After all, work must develop from the economic analysis made by the FSB. In taking into account both the flow of funds and the sector balance sheets, the FSB has developed a *relational mapping* of financial transactions (so-called "macro-mapping") that is able to indicate the points of interconnection of the shadow banking system with the regulated capital market (and, therefore, the mechanisms of transmission from the first to the second of the risks generated outside of the scope of supervision).[10]

It should be noted, moreover, that an option for a two-phase research methodology was the basis of this mapping. In this context, the FSB draw the line of the shadow banking system; and this is the border of the industry under a risk-focused monitoring (which should be complementary to the banking supervision).

In other words, according to the procedure followed by the FSB, the selection of subjects and operations here (to be included within the definition of the system under consideration) will have to take into account the need for a closer look at the cases that increase the levels of systemic risk and at those carrying out regulatory arbitrage in conflict with the purposes of supervisory public intervention.

This is a point of view that also allows focus to turn to the regulatory impact of EU rules on the operations that are carried out outside the traditional banking system. This approach leads to the obvious consequence of considering that certain transactions (of market-based financing) are somehow related to the desire to evade the safeguards of the supervision (put in place to ensure the regular operation of the industry).

In this context, the analysis of the shadow banking system cannot not exhaust its significance in reference to the quantitative assessment of the impact of the phenomenon on the possibility of financing the real economy, on translation of risks, and, ultimately, on the circulation of wealth. An analysis is needed that extends itself to impact the legal examination of the option of allowing the transfer of money under specific conditions (which are unachievable through the direct involvement of a bank).

However, it is clear that the common definition of a shadow banking system comes from an observation of the phenomenon carried out in order to "assess and mitigate systemic risks posed by other shadow banking entities." This purpose, in fact, is reconnected to the need to reduce the "spill-over effect between the regular banking system and the shadow banking system" and consequently to the need to regulate—in more conservative ways—the securitizations (and the incentives associated with them), in order to avoid the pro-cyclical effects of certain "secured financing contracts" (i.e. repos (repurchasing agreements) and securities lending) and, therefore, reduce the subjection of so-called money market funds (MMFs) to the phenomena of "runs."[11]

The same has to be said for the definition of the phenomenon formulated by the IMF, which reduces the essence of shadow banking to the ensemble of "many financial institutions that act like banks [but] are not supervised like banks."[12] Also, in this case, the linearity of the approach appears evident, to which the IMF correlates the detection of a "form of regulatory arbitrage" and, at the same time, the ability to perform an "important financial intermediation function distinct from those performed by banks and capital markets, as confirmed by the its continued growth."[13] Therefore, an inclusive approach can be highlighted to include the shadow banking system in the global financial system, unless providing appropriate forms of regulation and control.

The G20 also leans towards this conclusion when they highlight—in dealing with the problem of the regulation of the global financial system—the need to "address shadow banking risk," even for the purpose of increasing the safety of the cross-border financial transactions (also in derivatives), and thus the overall resilience of the whole capital market (according to the warning that "the prospects for global economic growth to strengthen in 2014 but remain vigilant in the face of important global risks and vulnerabilities").[14]

1.3 The directions of certain central banks

Consistent with the statements discussed in section 1.2 are the evaluations of the German Bundesbank. It suggests the need for the strengthening of the oversight on the shadow banking system. The Bundesbank has analyzed the implications of this phenomenon on the monetary policy, focusing on its effects on the euro zone. Therefore, these findings will be taken into account with the will to follow the stability in the European Monetary Union (EMU) internal market.

Adopting a similar perspective to that of the FED, the Bundesbank claims that the shadow banking system "can impact on the effectiveness of monetary policy measures."[15] Hence, there is the possibility that the growing importance of the first could change the mechanisms of transmission of the second. Consequently, it highlights the need to extend the scope of supervision, also in view of the known objective of maintaining price stability (to which is oriented the action of the ECB, art. 127, Tr. FUE).[16]

On this point, the analysis model of the Bank of Italy is very interesting. This model, in a macroeconomic perspective, evaluates how the "commercial banks can offload risky loans onto a shadow banking sector." The approach shows the assessment of "interactions and spillover effects," which implies that "high leverage in the shadow banking system heightens the economy's vulnerability to aggregate disturbances."[17]

The position of the Bank of England also is in line with this approach. In sharing the conclusion provided by the Commission, it promotes a "possible concrete policy agenda."[18] However, the British authority made a reformulation of the definition, by describing the shadow banking system as "credit intermediation, involving leverage and maturity transformation, that occurs outside or partly outside the banking system" (and then modifying a previous interpretation in which there was also a reference to "monetary services").[19] It is, therefore, possible to find a convergence of the phenomenon and the non-bank financial sector, with the effect of excluding—from this system—the hedge funds that do not perform intermediation activities of the type indicated above.

However, the British approach to the shadow banking system highlights a number of particularities other than those examined so far. This applies, in particular, to the option to also trace back, to the phenomenon under observation, operations that, at least in part, involve commercial banks (which provide the necessary resources for the implementation of financing through credit lines granted to the vehicles established for the operations).

It is clear that this approach has specific consequences. According to it, we must include within the system the operations that do not create new liquidity, but use the traditional banking system. As the Bank of England has acutely observed, in this case, "there is a real bank in the shadows."[20]

Consequently, the actions of the FED and its "quantitative easing program"—begun in 2008 and closed in October 2014—must be taken into account. The choice to buy mortgage-backed securities from banks was not only a clear expansionary monetary policy, but also a new form of open market operation able to connect the asset-backed security (ABS) to the money supply (by applying an originate-to-distribute business model).

Furthermore, other programs were designed by the FED to support the liquidity of financial institutions and improve safer conditions in financial markets. It is known that these actions have involved purchases of ABS and CDOs (and longer-term securities) aimed at managing (longer-term) interest rates and easing overall financial conditions, without taking into account the market-based nature of these assets. This is why these programs led to significant changes in the boundary lines of the shadow banking system—because (while many of the first crisis-related programs have expired or been closed within a short period of time) the FED continues to take actions in the *shadows* to fulfill its statutory objectives for monetary policy: maximum employment and price stability.[21]

So, this identification is essential for the purpose of implementing an effective system of supervision on the shadow banking system, to identify the cases in which its operations should be included in the scope of the traditional banking supervision system (and, therefore, subject to minimum capital adequacy in a perspective of safe and sound management of the assets in which are invested the savings). This leads to a definition that highlights the riskier profiles of the shadow banking system, which are able to produce effects that extend beyond the aforementioned scope and affect the capital market as a whole.

1.4 The routes of European institutions

At European level, a different approach prevails. Firstly, consideration must be given to the fact that the European Commission has launched a strategy aimed at programming an action plan to regulate the capital market. In this launch, there are the signs for a consultation process, designed to seek the (shadow banking) operators' input on matters affecting them. Obviously, the European Commission's goal is the definition of the scope of supervision, followed by guidelines to maintain stability (in the long run).

This strategy, to date, is synthesized in a specific proposal on the "transparency of securities financing transactions," which aims to avoid the presence of improper mingling between credit institutions and the shadow banking system—recommending also "to stop the biggest banks from engaging in proprietary trading and to give supervisors the power to require banks to separate those other risky trading activities from their deposit-taking business."[22]

Therefore, the contents of the European Commission's *Green Paper on the Shadow Banking* must be taken into consideration, in which this system is described as "an increasing area of non-bank credit activity... which has not been the prime focus of prudential regulation and supervision." The paper also observes that the "shadow banking performs important functions in the financial system... it creates additional sources of funding and offers investors alternatives to bank deposits. But it can also pose potential threats to long-term financial stability."[23]

Moreover, the analysis must include other findings of the European Commission, given that the Commission discovered how this system can be based on two intertwined pillars: the first is related to *entities* that, although not banks, carry out an activity traditionally reserved to the latter (i.e. "accepting funding with deposit-like characteristics; performing maturity and/or liquidity transformation; undergoing credit risk transfer; and, using direct or indirect financial leverage"); the second refers to the *operations* that can fulfill a function of investment and financing (i.e. securitization, securities lending and repurchase transactions, repo).[24]

However, the aforementioned conclusions of the European Commission shows the need for new supervision activities, in order to correct the market failures that occur in the *shadows*, and to manage the risks that go beyond the alternative credit channels (and could influence the whole capital market). To the identification of these risks corresponds, in fact, the possibility of adopting specific measures to avoid any degenerative perspective, with regard for the alternative and hedge fund managers (ruled by Directive 2011/61/EU, so-called AIFMD), and for the securitization transactions (ruled by (EU) Regulation 575/2013 and Directive 2013/36 EU).

This approach also explains the reason why the definition of the shadow banking system proposed by the European Commission in the Green Paper of 2012 has received the general approval of those who replied to the consultation.[25] Consider, then, a broad definition that, unlike the one used by the FSB, highlights the benefits and risks of the phenomenon under observation. This definition aims to establish a notion compatible with the purpose—expressed by the G20 in Cannes 2011—to develop the "recommendations" for the supervision of the operations in question.[26]

In this context, all the EU directives and regulations aimed to expand the scope of the European supervision (and then to reduce the shadows) must be considered. In particular, referring to the Directive 2014/65/EU and its assessment of the development of the internal capital market (in line with the goals set by the MiFID, Directive 2004/39/EC). Recent years have seen—as well as an increase in the number of investors who access it—an expansion of the tools and services offered therein (which also corresponds to an increase in operational complexity).[27]

In light of what has already been said here, it is easy to understand also that the expansive trends of the European financial supervision influence the role of the Regulation (EU) no. 648/2012, where the central counterparty (CCP) becomes a service provider interposed between the parties of a contract (traded on one or more financial markets) "as the buyer to every seller and the seller to every buyer."[28]

Furthermore, it is important to consider the provisions of Directive 2014/65/EU, which outlines a uniform system of supervision (on alternative funds) that, in addition to ensuring the quality of market transactions (at a micro-prudential level), pursues the macro-prudential goal to maintain the integrity and efficiency of the global financial system.[29]

However, it should be noted that the mentioned Directive 2014/65/EU does not innovate the regulation concerning access to the capital market by the sovereign states, central banks, and public bodies charged with or intervening in the management of public debt. In fact, it does not include these entities in its scope (art. 2, exemptions).[30] This option has been subject to a specific assessment by the Commission in the amendment of Regulation (EU) no. 648/2012 (with particular regard to derivatives traded over-the-counter); and the following choice—of not limiting the powers of the aforesaid entities in the exercise of their duties (considered of common interest)—relates to the results of this assessment. But these results do not cast doubts that the public administration, in order to find resources, will end up with an attempt to access the shadow banking system (and, therefore, operate in an environment conditioned by insufficient levels of information transparency or inefficiencies in the process of the formation of prices).[31]

Undoubtedly, these provisions draw a border to the *shadows* consistent with the needs of the current financial market set up. According to the new architecture of the ESFS, we can identify the attempt (of EU regulators) to extend the scope of supervision to the new generation of "organized trading systems," and then to apply—to *people* who operate in it—certain obligations to prevent the lack of regulation from destroying wealth and, in this way, jeopardize the efficient and orderly capital circulation.

At this stage of the analysis, it should be clear that the definition of the shadow banking system is conditioned by the measures introduced to mitigate the effects of the financial crisis (and, therefore, by the changes to regulation that effect the functioning and transparency of the capital markets). In particular, we must take into account the European regulatory option to place under supervision the off-exchange trading of certain instruments (OTC), according to specific rules that—in addition to pursuing the goals of increasing transparency, protecting investors, and strengthening confidence—attribute to the authorities more intervention powers (and, therefore, lead the operations in question within the traditional surveillance mechanisms).

According to the previous considerations, the contents of the most recent EU rules go to the strengthening of the regulatory framework and, hence, it reduces freedom within the shadow banking system. Consequently, it can easily be understood why these rules mark a clear distinction (or rather, division) between the shadow banking system and the supervised credit institutions (as defined by Directive 2013/36/EU). It is also clear that this distinction becomes even clearer if linked to the prospect of a "single European rulebook," applicable to all financial institutions within the internal market, intended by the European Council to support the future European supervisory architecture.

Therefore, the aforementioned Directives no. 2013/36/EU and no. 2014/65/EU and the (EU) Regulation no. 600/2014 anticipate a supervisory framework that—at a European level—will delimit the shadow banking system (being applicable to banks, investment companies, regulated markets, service providers of data communication, and companies from third-party countries who perform services or investment activities in the European Union). This approach is in line with the tendency of the EU authorities to minimize, where possible, the discretion left to the Member States for the transposition of European legislation (in the field of financial services). Therefore, the national differences do not appear destined to trigger forms of regulatory competition, nor to allow—for the competent authorities—the possibility to delay the control of the interconnections between the shadow banking entities and the supervised ones.[32]

1.5 The path of emerging countries

Certain emerging countries endorse the aforesaid logics. Here I refer to the analysis of the Reserve Bank of India, which has reconnected the shadow banking system to the "bank-like functions performed by entities outside the regular banking system."[33] Once again this is a definition that highlights two of the most important assumptions of this system: the similarity to the banking activity (with reference to the activities of maturity transformation, undertaking credit risk transfer, and using direct or indirect financial leverage) and the specificity of the operations through which the capital circulate (i.e. securitization, securities lending, and repo transactions).[34]

It is imperative to consider the empirical evidence, from the financial press, that in China, "every time regulators curb one form of non-bank lending, another begins to grow."[35] It is particularly important to take into account that while China's economy is slowing, shadow banking entities raise money from businesses and individuals by offering returns as high as 10 per cent (whereas the government imposes a low cap for interest rates on bank deposits).[36]

In this context, it is clear that the volume of the shadow banking operations can influence the development of national economic policies, which could not reach the goals set by the governments. Consequently, we must take into account that the shadow banking system intermediates a significant amount of transactions in these emerging economies, which affects "nearly one-third of aggregate financing in the world's second-biggest economy," that is, the Chinese market.[37]

1.6 The interpretations of the phenomenon

The analysis of the alternative form of capital circulation conducted by the Federal Reserve Bank of New York appears to be oriented towards a different classification of the shadow banking system. At first glance, it seems that the observation of this system is influenced by the observer; and, indeed, the FED moves away from a monetary perspective, assuming that "shadow banks are financial intermediaries that conduct maturity, credit, and liquidity transformation without explicit access to central bank liquidity or public sector credit guarantees."[38]

This is, therefore, an analysis that attaches primary importance to the institutional features of the so-called shadow banks, to which is given a central economic role in relation to the dynamics of the circulation of relevant cash flows. Hence, the boundaries of the shadow banking system include all the "sources of funding for credit by converting opaque, risky, long-term assets into money-like, short-term liabilities," being understood that "credit creation through maturity, credit, and liquidity transformation can significantly reduce the cost of credit relative to direct lending."[39]

Moreover, in this American interpretation, the distinction between the shadow banking system and the traditional banking channels had been founded on the possibility to receive funds from the FED's discount window or other tools of monetary policies or institutional guarantees (like the Federal Deposit Insurance). It is not possible to consider all the entities linked to the emergency liquidity facilities, which should help only traditional banks. Interventions that only save banks operating with shadow entities would lead to a liquidity crisis (or rather, the lack of credit resources resulting from a crunch of the shadow banking system), by safeguarding the liquidity of the institutions that have signed the securities in default (or allowing them to activate bank refinancing operations to support companies that had previously enabled a form of market-based financing).[40]

In other words, the abovementioned boundaries can be traced along the channels of monetary policy transmission, which are included solely within the traditional banking system. Furthermore, only the private financing channels use the shadow banking system, and so the liquidity and solvency of the operators influences the functioning of the whole system (which acts without public backstops).[41]

The focus, then, must be on the regulation of the relationships between any national (or sovra-national in the case of ECB) monetary authority and the subjects of the shadow banking system. This focus leads to the option—studied by the European Commission—of orientating any form of market-based financing towards the centralization of payments (and, therefore, imposing the intervention of clearing service providers). Indeed, there is the possibility, for any monetary authority, to refinance a central counterparty in crisis. This possibility directly connects the monetary activities to the market-based financing, with obvious effects on the social welfare (arising from the additional placing of liquidity). This can be a solution in cases where the private system is not able to refinance the shadow banking system.[42]

According to this approach, in extending the boundaries of the phenomenon, the following can also be included:

- all the operations that do not benefit from the possibility of taking advantage of backstops or other forms of last-resort support; and
- all forms of credit intermediation that are not "officially enhanced" by central banks (or other guarantee schemes from public institutions).

Consideration should, therefore, also be given to the possibility of tracing back to the *shadows* the securities lending activities—of insurance companies, pension (or health) funds, and other managers of "other people's money"—which do not have access to financing operations from the aforementioned public authorities. Undoubtedly, there are numerous supervised entities—enabled to manage money collected according to insurance or financial schemes, and invested in accordance with the applicable rules that have been established to safeguard the creditworthiness of such subjects. However, these entities can access the shadow banking system both as lenders (by investing their portfolio in securities that refer to shadow operations) and borrowers (searching in such a system for the resources to finance their leverage policies).

This has resulted in a specific link between the shadow banking system and the traditional credit and financial institutions. These institutions, therefore, are exposed to the risks of the alternative intermediation channels. Hence, the public policies shall take into account the *shadow dynamics* also in the regulation of insurance, finance, and social security industries. There is no doubt that their investment policies (that characterize the management of portfolios, of technical reserves, and of the assets placed as coverage)—together with the use of leverage—will have to provide countercyclical buffers and other controls (designed to prevent the critical issues arising outside the scope of application of the supervision shifting towards the sectors just mentioned).

We reach, then, a specific conclusion. According to the American approach, the definition of a shadow banking system includes certain specific affairs: "(1) the government-sponsored shadow banking sub-system; (2) the 'internal' shadow banking sub-system; and (3) the 'external' shadow banking sub-system."[43] The following chapters will look at the specification of the characteristics of the aforementioned subsets, but it is useful to point out here the distinctive elements of the subjects who organize the operations.

In other words, the interaction between the form of market-based financing under consideration and the diffusion of the high frequency trading and algorithmic trading should be remembered. In more recent times, the technology of trading—of securities that give content to the shadow banking system—has undergone a profound change that determines its automatic configuration. And this should be noted with regard to purchase orders and sales when, in such transactions, human intervention is negligible (if not non-existent).

This poses an interesting problem, given that, in a single period, any operator activates a plurality of contemporary and identical orders (to then conclude only with the most convenient), or it proceeds to the automatic fragmentation of a single order (mediating the results and minimizing the risk of incurring a misleading price with little benefit). It goes without saying that such ways of negotiation amplify the volumes, and thus affect the production of information and the dynamics of prices, making it necessary to apply a new type of supervising requirement, aimed at purifying market data from the just mentioned *dystonia*.[44]

Central—in this case—is the supervisory authorities' knowledge of the formulas used by trading systems to analyze the "cognitive inputs" (i.e. data or signals of the market) and the algorithms by which the "output decisions" (i.e. the orders processed within a very short time, in response to the analysis of price trends and other market variables) are determined.

In general, it can be said that the development of the trading technology has both improved the market functioning and simplified the execution duties. However, with regard to the shadow banking system, we can observe ambivalent effects. On the one hand, there is a wider participation in operations, the levels of liquidity increase and the transaction fees reduce (and, because of the high frequency of trading, the differentials and volatility of prices reduce in the short term). On the other hand, there are triggered risks associated with the overload of information systems (of the trading venues managers), to which may also correspond the generation of erroneous orders and the effect of undermining the smooth functioning of the capital market.

1.7 The different outcomes of the monetary and supervisory perspectives

Drawing an initial conclusion, we can say that the definitions of shadow banking system currently in use appear to be formulated on the basis of prudential (that of the FSB) or monetary considerations (the FED's). However, any of them provides fully functional interpretations for the implementation of a system of supervision based on a careful and constant monitoring of the phenomenon.

It will, then, be necessary to pay attention to the quality of the assets underlying the securities issued in financing transactions, the efficiency of the circulation of the latter (and, therefore, the transparency of the process of price formation), and the security of the payment system (also promoting the intervention of a central counterparty able to eliminate the risk of insolvency).

In the end, the goal is alignment of the public supervision to the dynamics of financing that characterize the system under consideration, in order to regulate and control its development and the ability to promote higher levels of wealth.

As a consequence, if there is a setting favorable to the affirmation of the shadow banking system, then the public intervention shall not penalize those who wish to take advantage of capital markets that are alternative to banking channels (without, however, affecting the need for a full application of the principle of equality—applying similar rules to activities that present similar risk). On the contrary, it shall protect the price stability and the savings.[45]

Obviously, the substitutability of the two financing systems puts shadow entities and banks in direct competition, with implications in terms of supervision and, therefore, in the identification of transactions and of risks that must be subject to (prudential) regulation and (public) control.

From a competitive perspective, a broad interpretation of the shadow banking system suggests assessments, with regard both to the support of an advanced capitalist system and to its propulsive effects in the development of corporate finance (and, therefore, to an improvement of the financing practices made by the industrial sector).

In other words, the shadow banking system consists of various operations and—in line with the approach adopted by global regulators—it requires a more effective intervention of supervision; because this system is not able to correct by itself all the market failures (and then ensure a competitive playing field between alternative financing systems). Furthermore, the system is not able to expel those subjects who adopt opportunistic behavior (i.e. those in which the goal of a predatory profit prevails over the intent to increase the supply of the forms of intermediation and, therefore, to multiply the wealth in circulation).

This corresponds—even in the shadow banking system—to the widespread preference for higher levels of transparency and liquidity, which make it possible to move available resources towards the subjects that are able to use them in the most convenient ways.

Therefore, it seems preferable to start from a definition of the shadow banking system that does not qualify it as a space of irregularities, but rather as an *alternative meeting point for demand and supply of capitals*, in which there are opportunities different from the traditional ones (because they are not subject to the parameters—of capital adequacy and risk control—introduced by the Basel Accords).

1.8 The boundaries of the shadow banking system

1.8.1 Money laundering, tax evasion, and other forms of "black market"

In the light of what has been discussed in this chapter so far, the lawfulness of the shadow banking system—which, in its basic components, appears as a mechanism of capital circulation (and this, it is worth repeating, in ways that can ensure the maturity transformation, the reshaping of the risks, and the management of financial leverage)—can be clearly understood.

It goes without saying that this system is clearly distinct from the traditional banking channels, which represent the outer limit of the shadow credit intermediation process—as it is, on the one hand, subtracted from the prudential supervision (of public authorities) and, on the other hand, excluded from the monetary transactions (by central banks).

That said, it should be noted that the phenomenon under observation does not connect finance to crime. Even in the shadows, in fact, there is no possibility to perform operations in order to:

- pursue the goal of transferring the benefits of illegal activities towards civil society (i.e. recycling);
- allow the money transfer to persons who commit, attempt to commit, or support terrorist acts (i.e. counter-terrorism);
- hide or reduce the tax basis of an individual or a company belonging to a particular system (i.e. tax evasion); or
- fuel the illegal trade of goods and services (i.e. the black market).[46]

With particular reference to anti-money laundering and anti-terrorism, it should be noted that a group of international and European standards have been set in the fight against the illegal trafficking of drugs and psychotropic substances (Vienna Convention, 1988), the proceeds of crime (Strasbourg, 1990), the financing of terrorism (New York, 1999), and corruption (Palermo, 2000). In addition, one must consider the Warsaw Convention that, after the events of the early 2000s, marked a major step forward, indicating the measures to be taken at national level to implement fully the International Convention for Suppression of the Financing of Terrorism, adopted by the General Assembly of United Nations on December 9, 1999 in New York (cited earlier).

It is clear, therefore, that the use of the shadow banking system for these purposes of laundering (the proceeds of crime) or financing (the terrorism) is contrary to the provisions of the aforementioned international standards (and their national implementing legislations). In other words, any prohibited action cannot be carried out either within the traditional circuits or the shadow credit intermediation process.

There is no doubt that the lack of transparency that, as has been anticipated, characterizes the shadow banking system may induce some individuals to attempt to hide "in these fogs" the conversion or the transfer of goods that come—directly or indirectly—from criminal activity. In particular, I refer to the will to *camouflage* the illicit origin of money or to evade the legal consequences of one's actions.

We are aware of the risk that money laundering or terrorist financing may—in some way—take advantage of the alternative channels of intermediation of capital under consideration. In facing this risk, then, there is a need to trace the money and financial instruments (within the shadow credit intermediation process). This is a requirement that, as shall be seen, will be the basis of the intention to strengthen the supervision on the shadow banking system and, therefore, may be satisfied by the implementation of new systems of data collection, as well as the sharing of these with the police authorities.

Concluding on this point, the restrictive tendency that—in order to combat money laundering and terrorism financing—reduces the cash circulation must be taken into account. Before the withdrawal of the $10,000 note (issued by the Bank of Singapore) and the restriction for the use of the €500 note (by the ECB), we can find a generalized option for *traceable* forms of payment that interact not only with the common practice of settlement of transactions based on the real economy, but also with the process of financialization of enterprises.

In limiting the use of cash (and, therefore, entrusting the incomes at the times set by payment institutions) and in reducing the possibility of using alternative methods of financing (among which is the practice of transferring bank checks), new "financial needs" are generated and, consequently, entrepreneurs are faced with the necessity to undertake a critical review of their funding sources (allocated as fixed assets). Therefore, they must proceed in search of new tools of corporate finance that are able to meet the needs generated by the anti-money laundering regulation.

In other words, we can say that if, on the one hand, the fight against the financing of illegal activities (and then the limitation of cash circulation or the transferability of certain securities) marks a boundary of the shadow banking system (by prohibiting certain risky and opaque transactions that gave content to the system), on the other hand, it pushes executives towards the exploitation of any other financing possibilities offered by the capital market and, therefore, towards the alternative channels of capital circulation (see Table 1.3).

The issues raised by operations in the shadow banking system are different, with the sole purpose of reducing the effective tax rate. Indeed, this is not a form of advanced financing, but a tax evasion that is

Table 1.3 Boundaries

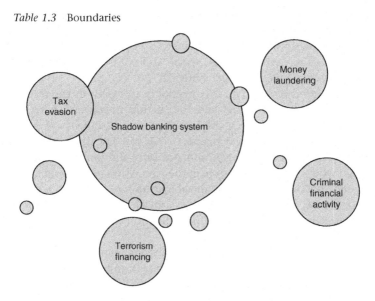

the end-point of a global degeneration. This is, in other words, an unacceptable attempt to avoid the social obligations imposed by the contemporary social-democratic countries.[47]

In particular, this effort often results in actions designed to hide the profits of the activities performed in the intermediate stages of the shadow credit intermediation process or to refer them to transactions that take place in the so-called "tax havens." Moreover, such actions may affect both the investments made through foreign vehicles (disguised as foreign in order to send the revenues to countries that carry a minimum tax levy) and the debts incurred for the sole purpose of generating a charge (for interests) deductible in countries with high tax burden. Further, the more complex cases consist of the joint performance of these two types of actions (where the same economic entity had contracted a loan in a country and subscribed to the related CDOs in another).

It goes without saying that such practices cannot be implemented within the regulated market, or in the shadow banking system. It is their nature that is in opposition to the social-democratic system, and it makes them incompatible with the global financial system, because of the relationship that reconnects the latter to the "general utility" and, therefore, to the objective of sustainable maximization of social welfare.

Therefore, one of the challenges that the process of strengthening the supervision on the shadow banking system will have to face can be identified: the provision of an "open finance," compatible with an economy conscious of the environmental, social, and political context (which is reflected by the applicable legal system).[48]

From another perspective, it is necessary to dwell on the difference between the financial crimes and the shadow operations, where only the transactions that breach the law must be prosecuted and, as such, excluded from the financial market.[49] We can say that, in this case, the transaction itself is illegal, and not only the financial operations through which it is carried out (e.g. the circulation of ABS and CDO) or the assets that are its object (financial instruments or money). We must consider these transactions as part of a black market, which is related to the "underground economy" (including numerous crimes, from drug trafficking to terrorism).

The same has to be said for the transactions relating to the so-called "unreported economy," that is, the business conducted outside of the ordinary rules of accounting and traceability.

1.8.2 Prohibited shadow operations

The fact that the structure of a shadow operation can breach certain legislative prohibitions, especially those provided to reserve banking to credit institutions, cannot be excluded. In particular, I refer to the case in which:

- the loans at the basis of a transaction are sold from one bank to a vehicle;
- the latter securitize the loans in short-term securities (so-called ABCP); and
- the originating bank subscribes the ABCP (on a continuous basis, through an operation of roll over).[50]

These kinds of operations should be prohibited by a general *anti-avoidance* rule, such as to declare unenforceable to the supervising authority any transaction that has no valid economic reasons, when such a transaction aims to circumvent prudential rules and, therefore, intends to achieve a more advantageous result in the computation of the Basel parameters.

In the European context, the forthcoming adoption of the "Single Rulebook" offers the possibility to insert such a norm in this set of harmonized prudential rules. Moreover, in perspective of the EBU, an anti-avoidance rule will prevent the problems arising from the "national regulatory patchwork" (currently in force in the EMU), which has shown its limits during the crisis, leading to legal uncertainty and then enabling institutions to exploit regulatory loopholes that distort competition.[51]

In other words, the process of reshaping the EU (supervisory and regulatory) system should give the competent authorities more effective powers (than the ones aimed to mitigate the risks arising from securitization transactions). These provisions shall guarantee that banks will be evaluated and assessed according to appropriate policies and procedures, as specified in art. 82 of Directive 2013/36/EU.

In light of what has been discussed so far, it is clear that the "Single Rulebook" will have to provide a wider intervention, in order to ensure that the "economic substance" of any transaction produced by a bank is not the aforementioned circumvention of the parameters of prudential supervision and, therefore, that its accounting treatment will be consistent with actual levels of risk.

To summarize: not only the breaches of law but also the attempts to escape from the market rules must be placed outside the perimeter of the shadow banking system. Both these actions, in fact, try to achieve the opportunistic goal of unfair individual profit.

On the contrary, it is necessary to trace back to the shadow banking system (and to other lawful forms of alternative market-based financing) the operations that have been realized without the intervention of a credit institution (or any other supervised financial firm) and, therefore, without the involvement of the funds collected from its customers.

Obviously, these operations avoid some of the safeguards imposed by the Basel Accords. There is no doubt, in fact, that the shadow credit intermediation process does not fall within the actual scope of the regulation of capital adequacy.

Thus, this process is not a way to breach the rule, and it cannot be defined by the components at the basis of its phases, in which the realization of various shadow banking operations, by variegated shadow banking entities, connect subjects in *deficit* with subjects in *surplus* (and, in this way, promote new methods of financing the real economy by using alternative undertakings for investment. The result is, therefore, that shadow banking can be qualified as a system that is efficient in circulating capitals, according to economic parameters different from those that characterize the banking industry (in terms of risk taking and return's profitability).

2
The Shadow Banking System as an Alternative Source of Liquidity

This chapter considers the key economic drivers of shadow banking. It begins by examining the efficiencies of this system, that are the rationale for the bundling of activities that we define as market-based financing. The chapter goes on to take into account the market failures amplified by the shadows, focusing on asymmetric information, lack of transparency, and market instability.

The chapter will clarify that the current legal framework allows a lawful use of business freedom, but not a means of escape from banking supervision. In this context, I address the risks related to the global nature of the shadow credit intermediation process and to its cross border transactions.

2.1 The economic determinants of the shadow banking system

It is important to develop the legal analysis of the *shadow preferences*, which were introduced in a preliminary way by defining the scope of the analysis. The subject of this chapter is the utility of the shadow banking system, and its contribution to the efficient development of the global financial network as a competitive market (alternative to the banking system in financing the real economy).

We are aware that these economic features have already been highlighted by numerous economic analyses.[1] Undoubtedly, it should also be considered that the economic determinants of the shadow credit intermediation process are closely connected with the demand for capital and the supply of "safe, short-term, and liquid 'money-like' claims to invest...large cash balances," as well as the offering, by traditional banks, of instruments "that use securitized safe and long-term 'AAA' assets to attract repo funding (from institutional cash pools directly or money market funds) and boost leverage."[2]

In analyzing the economic determinants, however, it is still convenient to link them to the definition of a shadow banking system in which "shadow does not necessarily mean dark and sinister."[3] As was highlighted in Chapter 1, we are in the presence of a system in which a process takes place that can provide further (and additional) funding than that allowed by banking (subject to the constraints of prudential regulation); together with the possibility of increasing the resources available to meet the needs of the real economy (without undermining the global financial stability).

Significant, in this regard, is the fact that the supply of capital (accomplished this way) requires lower (operating) expenses, related to the flexibility of the organizational structure of the entities that operate in the shadow system (on which less transaction costs are imposed).[4] Even on the capital demand side, then, individual firms may have reasons of convenience for accessing the shadow banking system. Here I am referring, in particular, to the achievement of specific benefits associated with the diversification of funding sources (with respect to both the counterparties and the instruments).

Obviously, corresponding to such a structure is an increase in the levels of financialization of the capital structure of the entrepreneurs who access the shadow banking system to draw resources to invest in their business. It seems, therefore, necessary that the debtor enterprise would implement the governance with new business functions, responsible for the monitoring of the funding sources and, therefore, for the measurement and the management of the risks associated with debt obligations that are intended to be securitized and used in complex transactions (which give content to the shadow credit intermediation process).

One should consider that the absence of regulation and control allows companies to negotiate the riskiest financing through this alternative market. It is also possible for a debtor in difficulty to access the shadow banking system in order to draw resources to repay loans previously taken from the traditional banking system. This occurs not only because the company manages to find the most favorable conditions, but also because it could be illiquid and, therefore, presents a high-risk profile (too high to obtain the renewal of bank loans).[5]

From another perspective, the economic determinants of the phenomenon also relate to the benefits that it reserves for the subjects in *surplus*. There is no doubt, in fact, that in the shadow banking system there are investment opportunities of particular interest, both in terms of risk remuneration, and of safety and liquidity of the issued securities.

Therefore, it can be said that the alternative nature of the shadow banking system (to the banking sector) determines both a lower incidence of costs (incurred in similar conditions by banks), and the diversity of the warranty conditions (due, among other things, to the absence of procedures for crisis management of the issuer). From here, higher remunerations are offered (for the same risk) and securities issued are negotiated at the end of less reliable operations (because they are made in a context free from the traditional regulatory protection; see Table 2.1).[6]

This is only another way of saying that the search for an environment characterized by lower transaction costs is one of the operators' incentives to get involved with the shadow banking system. But it appears to be in line with the essence of intermediation (which—as is known—tries to put people who would not otherwise meet in contact with each other and, therefore, would not engage the deal).[7]

We must also refer to the type of financial instrument chosen for the realization of shadow banking operations. On this point, attention should be paid to the market orientation towards short-term securities—also highly rated (as guaranteed by subjects with primary standing)—and, therefore, to the presence of a demand for "safe, liquid assets from corporations and asset managers," exceeding the absorption capacity of the banking system.[8]

Table 2.1 Determinants of the shadow banking system

- Increase of capital supply
- Lower incidence of transactional costs
- Wider range of warranties conditions
- Possibility of high risk taking
- Higher remuneration for risk taking

This is the conclusion reached by the analysis promoted by the IMF. It has encountered the inability of banks to absorb the supply of liquidity resulting from cash pooling operations of the aforementioned subjects.[9] Furthermore, there is the fact that the administration of cash pools is, usually, entrusted to offshore subjects (or, at least, outsiders to the legal system of the parent corporation), which may prefer operational forms that are realized on international markets to a bilateral relationship with one or more banks. This explains the IMF's conclusion that at the basis of the growth of the shadow banking system there was (and still is) an "accommodating cash-pools' demand for safe, short-term liquid assets in volumes larger than those (inelastically) provided by short-term government debt."[10]

It is the "mismatch" between these (i.e. private demand and supply of short-term sovereign debt) that places the incentives that will lead to an offer of CDOs that apparently responds to the aforementioned liquidity needs. It seems, therefore, possible to conclude that it was the demand for cash pools to orient *spontaneously* towards the securities of the shadow banking system, which (especially during the crisis) were perceived as *perfect substitutes* for bank deposits (because of the attested security qualities).

More generally, it is necessary to observe the conditions of competition that qualify the market structure of shadow banking in comparison to that of traditional banking. There is no doubt, in fact, that the financial regulation allows entities that want to offer their liquidity (and, therefore, ask for short-term investment instruments) to operate—alternatively or jointly—in both markets. There are not, in fact, prudential safeguards designed to prevent the business functions appointed to manage cash pools having access to one and/or the other system.[11]

It should be carefully considered whether the coexistence of these two systems has increased the overall efficiency of capital circulation (also through banks' diminishing price for brokerage services, i.e. rates). Variable effects have been produced, that are not always in line with the objectives of stability and proper functioning through government intervention.[12] This applies, in particular, to the increase in yields and, therefore, to the reduction in intermediation margins of banks (with obvious negative impact on profits and capital requirements of the latter).

That said, the substantial convergence of the expectations of financial traders (i.e. related to the ability to derive gains from the circulation of capital) should be considered. The choices of products offered by banks or of CDOs are based on these. Therefore, in the determination of investment options, prices—and their ability to orient the operators towards one or the other system for the rationally profitable resources available to them—will have specific importance.

It is evident, therefore, that there is a substantial competition, marked by trade-off between degrees of freedom, safety, costs, and performances.

In light of all of this, the multiplicative ability of the shadow banking system is of particular interest, because—in it—financial operations are executed without the duty to provide a capital reserve or correlate the relative timing (for both active and passive commitments). We are, therefore, in the presence of a system that admits risky operations: either with the absence of leverage limits or the possibility that, in this context, money substitutes are generated (and, therefore, are realized deviations from monetary targets set by central banks).

The need for international authorities to monitor the shadow banking system is, therefore, clear. It should not be recognized as valid that the desire to achieve extra individual profits—higher than those realized in the banking market—can promote a fleeing of operators towards areas of freedom higher than those recognized by the financial supervision to credit intermediation.[13]

It is possible to draw a first conclusion regarding the fact that, in recent times, the general trend of economic conditions (low interest rates and widespread uncertainty) have led some subjects in surplus to use the shadow banking system in order to expand the operating volumes of their business (and so increasing the importance of the relevant industry). This has been matched by an increase in the supply of credit to the real economy. Obviously, this fact has resulted in new financing relationships (in terms of risk, maturity, territoriality, currency, etc.), with understandable positive effects in terms of the quality and quantity of economic resources.

There are clear reasons underlying the support for the development of the phenomenon under consideration. Nevertheless, this has led to the existence of a number of risks—the implications of which will be observed later—resulting from the conditions of information asymmetry, opacity, and instability that have characterized so far all the operations in question.[14]

In this context, however, the option for a market organized in accordance with less onerous rules appears—in principle—in line with the objectives of efficiency typical of the capitalist system. But, sometimes, this option is just the adhesion to multilateral trading systems (which are not always able to avoid shadows in the circulation of data and information). There are determined asymmetries, which are resolved to the detriment, first, of the weak contracting parties and, second, of the overall wealth of the market. It is clear, in fact, that in the absence of an adequate international legal order it does not appear possible to find guidelines that lead operators towards the maximization of social utility. Rather areas of freedom can be identified in which it is possible to carry out opportunistic behavior (at the expense of the other party or the entire community of reference).

2.2 Information asymmetries

In light of what has been discussed so far, it is evident that the growing importance of operations based on market mechanisms corresponds to an increase in the information required to identify the conditions of price and liquidity characterizing the circulation of wealth in the global financial system.

There are many consequences to overcoming centrality of the credit institutions (that took onto themselves the obligations arising both from the financing and taking deposits or other repayable funds from the public and granting credits for its own account), because the reliability of the relationship is linked to their correctness (and, therefore, the riskiness of the investment carried out is linked to their standing).

In the shadow banking system, moreover, there is not only a lack of banking institution, but there is also a proliferation of single companies that, for various reasons, contribute to the implementation of the operations in question (SPVs, advisors, arrangers, etc.). In addition, the signing of contracts (needed for the collection of savings and credit operations) is replaced by the subscription of securities resulting from the organization of a series of shadow banking operations (referable only indirectly to the subject in need of loans).[15]

Within these dynamics the transformation of credit, maturity, and liquidity is not realized within an institution, but in the market through the joint use of the tools of commercial paper (through a series of emissions) and of the warehousing of these (SPVs made to do it). These are the conditions that make the lack of information sharing possible among the parties of the credit process (consequently some operators have more data than others, being able to draw unfair advantage from this asymmetry).

We are, therefore, in the presence of conditions that—both at subjective (due to the absence of the bank) and objective (given the plurality of operations) levels—require the identification of the economic and legal content of any product offered in the shadow banking system.[16] Moreover, it is difficult to believe that such a complex market, as the system under observation, complies with the requirements of perfect information on the basis of the neoclassical models.[17]

Consequently the general information needs—based on rational decision-making processes—which relate to the effective weighting of investment choices have been identified; where the failure to satisfy these needs places doubts on the fact that, in the shadow banking system, the encounter between supply and demand is the result of cognitive activity properly aimed at the acquisition of exhaustive data.[18]

There is no doubt, in fact, that—during the financial crisis of 2007—the existing asymmetries (between the knowledge of those subjects who organized the shadow operations and investors) have resulted in a lack of risk perception, with obvious reduction effects on the roll over in emissions and an increase in financing costs.[19] In this context, only the most risky operations (for which a high interest rate was recognized) may be undertaken in competitive terms compared to the traditional banking circuit (so-called risk-insensitive funding).

It is clear that this condition, in determining a process of adverse selection of operations, leads to an increased level of risk of the shadow banking system,[20] because it is a situation that can be overcome by the registration of the vehicles and publication of the data relating to their assets (with regard both for the composition of the assets and for the structure of the guarantees). It would nullify, in doing so, the existing asymmetries between who arranges the transactions and who funds them (through the subscription of securities). The possible listing of ABS and CDOs in regulated markets could facilitate the flow of information, entrusting "trade repositories" (of new type) with the task of recording and storing the data in question.[21]

In other words, the overcoming of the abovementioned asymmetry is essential so that the shadow banking system can act as a perfect substitute for the traditional circuit.[22] So, this shows a double line of action, being able to predict both a restriction on the composition of the assets of the vehicles (by limiting the type of loans and securities, their maturity, and concentration indexes) and the containment of the maximum level of risk of the transaction (measured with regard to the presence of guarantee or positives ratings).

Concluding on this point, there is no doubt that the proper functioning of the market under observation can be ensured only by economic circuits characterized by fairness and transparency; that is, in order to ensure the possibility—for operators—of having full and effective understanding of the risk and the cost-effectiveness of each individual transaction.[23]

Therefore, it can be said that—in terms of effectiveness and efficiency—the presence of information asymmetries (regarding the reliability and quality of these transactions) raises issues that do not exhaust their significance in the context of the bilateral relationship, but stretch as far as to undermine the functioning of trading systems in question.[24] A disclosure failure creates a degenerative context in which it appears almost impossible to determine the quality of the products offered, with obvious consequences regarding unreliability of the data, dissemination of inaccurate news, or the perception—by operators—of a faulty informational system.[25]

Finally, the consequences of the absence of an intermediary and the enhancement of private autonomy in terms of participation freedom should be kept in mind. In fact, they devolve to the responsibility of each investor every assessment regarding the performance of the securities subscribed, as well as knowledge of the composition of the assets of the vehicles that issued them. This is a burdensome activity that, only in part, can be mitigated by the presence of a rating (of the product) or a third-party guarantee (able to provide an alternative performance in case of default).

2.3 Opacity, pro-cyclicality, and system instability

The consequentiality of the shadow credit intermediation process, apart from placing the entities that organize it into a condition of asymmetry, compared to those who act only in one of its phases, is an opacity of the operations' arrangements and of the entities' balance sheets. I refer, in this context, to the effects of the activity of warehousing that, as much as being intended for the unified management of a series of credit positions, realizes a concatenation of emissions that makes it difficult, even for the operators of the sector, to know the real performance of any business within the shadow banking system.

In particular, it should be noted that only in the first phase is a low level of transparency maintained, since the vehicles put in their assets, the loans (and, therefore, the credit relationships with companies) and a first set of financial instruments (such as ABCP). It is, therefore, still possible to identify—on the balance sheet of the entity—the necessary information to learn about the quality of the substance underlying the operation.[26]

Thereafter, the overall levels of opacity are increased by the intervention of other (second-level) vehicles that buy the securities issued by the first SPV (the aforementioned ABCP) and the simultaneous emission of other financial instruments (such as ABS). The same has to be said for the next phase in which the ABSs are again subscribed by another vehicle, which will then issue the CDOs, which—as mentioned already—are usually the final product of the shadow banking operations to which the funding function is assigned.

Hence, there is a sequential and complex dynamic, which more often than not presents an additional complication, because at each step the following vehicle diversifies its assets (purchasing, as anticipated, securities issued by various entities). In other words, a set of varied titles ABS or ABCP underlies each issuance of CDOs, and this creates confusion.

Ultimately, the "pooling of loans" first, and "structuring of ABS into CDOs," second, determine opacity levels that make it difficult to know the real economic content of the financial instruments in circulation. There is a problem about the possibility of consolidating and weighting the data (related to reliability and liquidity of the underlying loans) in each phase of the intermediation process. It is, therefore, necessary to provide a channel that feeds off the information necessary to monitor the development of the aforementioned process and, then, is able to consolidate it in reference to each issue of securities traded on the market.

The problem of opacity has obvious similarities with that of information asymmetries (in which, however, the unprepared subject is only on one side of the contract). It seems, therefore, possible to consider that—in this case—it is sufficient to introduce registration requirements (of vehicles and entities) and traceability (of the loans underlying the securities issued) to significantly reduce the uncertainties that undermine the shadow banking system. Conversely, it appears more difficult to enforce regulatory limits—to the composition of the SPVs' assets—based on the delimitation of the objects of the investment of the SPVs.

It seems possible to conclude that some of the structural opacity could be overcome through good use of "interlocking directorates," made in order. This will allow the exercise of powers of administration and control within the vehicle company by the entities that finance the operations: especially because of the limited value of the share capital within the SPVs and, therefore, the inapplicability of the known forms of shareholder activism.[27]

There is no doubt, in fact, that the relationships between investors and individuals who organize the shadow operations would benefit from the presence of a representative appointed by the investors in the administrative board of the vehicles of the first level. This would facilitate the control and monitoring of loans placed at the basis of the operation (especially in the presence of covenants). Furthermore, it would ensure certain investors of a position characterized by access to all the information necessary to overcome the aforementioned conditions of asymmetry and opacity (and this in conditions of advantage, but also of leadership compared to the other parties that, while having bought the financial instruments, are not in a position to appoint an administrator). It would create a market-leader figure that could influence other investors, both in the determination of prices and in the choice of an early exit or refinancing the transaction.

From another perspective, it is important to note the pro-cyclicality of shadow banking. The values of the shadow banking operations tend to sway in the same direction the main indicators of the economic cycle, sometimes increasing more than proportionally to the proxy. This is because of the presence of elements accounted at "fair value" in the financial reports and other accounting statements. I am referring to both the ABCP securities and the ABS. These financial instruments are issued on a market and, therefore, their price is the referential value recorded on the balance sheet of the vehicle that subscribes them (and this is repeated until the issuance of CDOs).

It might be more accurate, perhaps, to say that a fair and current assessment of the activities that make up the balance sheet of the SPVs incorporate the economic performance of the market, with the possibility of obtaining the issue of securities that are "structurally pro-cyclical." It is, in fact, the market price that, when taken as a parameter for the quantification of economic issues (ABS and CDOs), could lead to an expansion of the accounting value in proportion of the increase in prices for securities (and, conversely, reducing them at the time when the latter diminishes). This is, therefore, an operational mode that could help boost the width of the oscillations of the market's trends (in stark contrast with the principle of prudence that traditional banking activities are based on).

Further, the problems posed by the mark-to-market application to finance are well known; as are the solutions adopted from time to time by the regulators who tried to contain the pro-cyclical effect of the amplification of values. "Prudential filters," "countercyclical buffers," and other intervention on accounting regulation (of the financial vehicles) can mitigate the effects of fair value accounting principle, hence a clear prospect of involvement by the public authorities (whose task it is to avoid unjustified growth in the volume of the shadow banking system determining a rise in the supply of credit—squeezing the market share available to the traditional banking system—and, at the same time, interacting with the monetary mechanisms, undermining price stability).

Hence, other troubles arise because of the leveraging (and, therefore, the possibility that every vehicle can be financed through credit lines granted by banks or other privates). It allows the immediate exploitation of the increase in value of the securities in the portfolio, to which we can commensurate the amount of resources taken out as a loan (measured as the difference between total assets and the value of the securities issued). However, an operational methodology of this kind leads to an immediate reduction in resources available to the vehicle when the value of the securities declines, with obvious negative effects on the liquidity of the latter, on the need to proceed with a sale of securities to repay creditors and, consequently, on the value of the instruments issued by the same due to the sudden decrease in assets.

The position of those who have identified in the pro-cyclicality of the so-called "repo market" the origin of market failures that occur in the shadow banking system, has been pointed out in this regard. The evidence was accompanied by the proposal of specific rules aimed at introducing, on the one hand, "strict guidelines on collateral" and, on the other hand, "government-guaranteed insurance."[28]

Certain deeper perplexities—which may arise when it becomes clear that the informational asymmetries, the opacity of the operations, and the pro-cyclicality of the values are some of the key factors of the overall uncertainty of the shadow banking system—will prove that instability still does not find a solution in a suitable set of rules, given the deregulation of the financial market. Nor can this conclusion be avoided by arguing that the failure in correcting the abovementioned cases of bankruptcy determines the volatility of supply (of CDOs) and demand (for loans). In addition, there are repeated brokerage chains that expose each phase of the process—as well as the problems already mentioned—to "roll over" and "runs" risks (especially in cases where there is an excessive maturity transformation).

Obviously, the complexity of the aforementioned chains (and, therefore, the mixing of titles in each vehicle) increases the chances of contamination among seemingly separate compartments. These are, in fact, the risks which are not associated with any public intervention guarantee (of the type attributable to the lending of last resort), whence the danger that adverse effects of limited value interact with each other in a pro-cyclical dynamic and, thus, pose a risk to the overall resilience of the shadow banking system.

Therefore, it can be said that the "credit formula" underlying this system has elements of instability of peculiar importance, to which the systemic risks that can extend to financial markets are related.[29] This is not only because the process of maturity transformation presents the risks mentioned already (and is free from the traditional mechanisms of public guarantee), but also because the market participants can expand the volume of operations beyond the extent necessary to fund the credit to the real economy (i.e. loans) for the sole purpose of investing in securities (ABS and ABCP) that show increasing trends (and, therefore, to be able to generate capital gains in the short term, as accounted for at fair value).

2.4 Methods for classification of the phenomenon

At this point in the investigation it is worth noting the economic consequences of the definitions used by international bodies and authorities to describe the phenomenon, in order to understand what the consequence of a regulatory option that refers to the FED's or the FSB's approach might be. It is important to understand the dependence of the efficiency of a supervising system on changes in expectation, because it is this dependence that renders the effectiveness of the public intervention.

Two different ways of framing the shadow guide the interpreter in two directions (only partly divergent) towards the common need for monitoring the process and identifying the risks corresponds. On the one hand, the knowledge of the interconnections with the supervised system (and, therefore, the safety of the channels of contagion) and, on the other hand, the possibility of measuring the expansionary effects due to the offer of alternative credit (compared to the monetary policies).

Nevertheless, it must not be forgotten that, once again, the nature of the authority that observes the phenomenon seems to influence its definition (from the public law perspective). This can be helpful in the selection of regulatory interventions, in order to put each instrument (supervisory or monetary) in direct relation with its purpose (the stability of the market or of the prices). This leads to the relevance of the diversity of the ways of framing the phenomenon for the present purpose. It induces the policymaker to question him or herself about its validity, as happened during the consultation on the European

Commission's *Green Paper on Shadow Banking* (where they all agreed to the FSB definition).[30]

Moreover, the option for one or the other definition may correspond to the inclusion (or not) of some of the firms that operate in the supervised environment. There is, however, not much to be said about this *a priori*. The analysis will mainly depend upon the current choices of the relevant authorities, which appear able to take into account the operations that carry out "credit intermediation and maturity transformation," with obvious increased interest for "the issue of systemically relevant activities."[31]

For convenience of exposition, it shall be assumed in the following paragraphs that this is not an example of regulatory capture, but a free choice of intervention. The mechanics of placement set out already confirm the link between the system in question and the real economy, to which it provides funding and investment opportunities; conversely, it is excluded that it is of abstract, self-referential, and speculative character.[32] This does not prohibit thinking that some of the intermediate stages of this process are sometimes configured in redundant mode, that is, such as to be unnecessary compared to the cycle of brokerage that is desired to be accomplished (and preordained in order to increase the profits of the organizers).

Considering all this, it should be noted that both the approaches (of the FSB and the FED) highlight the worthiness of the interests involved. As a result, there is the problem of selecting the forms of regulation and control of the shadow banking system (regulation or money related). As Keynes wrote, "worldly wisdom teaches that it is better for reputation to fail conventionally than to succeed unconventionally."[33]

It seems useful in the face of this problem to classify the systems (shadow and traditional) in terms of service to the real economy. This leads to the question of whether the regulation can be entrusted only to the interventions of global regulators and, therefore, to the protection of the labile "soft law." From here, there is a move to a clear preference for an appeal to the paradigm of "hard law," which appears to be able to frame the phenomenon under observation within an appropriate regulatory framework, such as to be protected with the highest degree of effectiveness of the interests involved in this case.

Assume, for the moment, that the foregoing considerations confirm the sustainability of the shadow banking system. It is worthwhile, therefore, not only to attest the validity of the operations that achieve the fund raising activities (through the trading of securities in the financial market), but also to highlight that these operations assume systemic character and, taken jointly, give content to a process that is an efficient alternative to that in which bank intermediaries operate. It finds, therefore, confirmation that the relationship between the two systems of capital circulation is alternative and competitive (one market-based and the other bank-centralized). The same must be said with regard to the need to avoid undue admixture between them (having to prevent brokers from purchasing securities of the shadow system for the sole purpose of capital gains, and thus to improve their economic performance at the expense of the safety of the market).[34]

Now ordinary experience tells us, beyond doubt, that a situation in which there is a need for financing (within limits) is the normal case. Whilst firms try to resist a reduction of credit lines, it is not their practice to self-finance their own business whenever there is a rise in profits. That said, for proper placement of the shadow banking system, it is useful to consider the classifications carried out by the economic analysis, in order to divide the system into subsets that qualify for the effects that each of them is able to accomplish. This applies, in particular, to the distinction made between:

- government-sponsored shadow banking subsystem;
- internal shadow banking subsystem; and
- external shadow banking subsystem.[35]

With reference to the first subset, the choice to distinguish between the different operations that are carried out by government-sponsored enterprises (GSEs) according to an originate-to-distribute model of securitized credit intermediation appears to be agreeable. In the same way as banks, such persons shall undertake the transformation of maturities, financing on the market through the issuance of securities in the short, medium, or long term.

In emphasizing the point of departure from the assumed sustainability of the shadow banking, an important point of agreement must not be overlooked. It should be noted, then, that the operations have been distinguished into four main types:

1. Term loan warehousing provided to banks by the federal home loan banks (FHLBs).
2. Credit risk transfer and transformation through credit insurance (provided by Fannie Mae and Freddie Mac).
3. Originate-to-distribute securitization functions (provided for banks by Fannie Mae and Freddie Mac).

4. Maturity transformation conducted through the GSE retained portfolios (which operate not unlike SPVs).[36]

It means that, with the given assumptions, there is a subset in which the loans are not originated by those organizing the chain of operations, but instead come from traditional banks. Only the stages of loan processing and funding belong to the shadow entities, in which—as aforementioned—the activities of credit, maturity, and liquidity transformation are realized (in absence of public control and monetary backstops).

With regard to the transactions that give content to the second subset (the so-called internal shadow banking subsystem), different conclusions are reached. In dealing with certain private banks' option to abandon the model originate-to-hold, the organization of shadow operations is able to increase their levels of return on equity (RoE). In the shadows, in fact, may sit the transactions triggered by traditional intermediaries, which, first, provide the loans in view of their storage in a vehicle and, second, use a securitization that is able to issue ABCPs or, directly, ABSs. Granted, then, that the propensity to invest in these securities is a fairly stable function, this is the chance to make an immediate reclassification of the assets of the banks (which sees the substitution of securities for loans) and, eventually, their allocation in an off-balance sheet asset management vehicle (so-called OTD, to the extent that this is still possible as a result of changes introduced by Basel III rules).[37]

This is especially the case where there are only short periods in view, but this way of acting innovates the banking activities of supervised subjects, in which notice is not given to any of the traditional characteristics of credit-risk intensive, deposit-funded, or spread-based process. However, apart from short period changes, other and different typologies can be identified: market risk-intensive, wholesale-funded, and fee-based processes. This will lead, as a rule, to a change that brings—at least in part—an adaptation of the system of prudential supervision (which, as it will be seen, could address the problematic transition from the originate-to-hold model to the OTD one, updating the capital adequacy rules laid down by the Basel Accords).

It follows, therefore, that, in the OTD model, banking is not parallel to the intermediation that takes place in the shadow banking system, but it is a part of it (in respect of which it stands in sequential terms and, therefore, complementary). Hence, the new levels of banking efficiency and profitability are related to the liquidity of the markets, with the result that any imbalances of the past have direct effects on the performance of banks (that have adopted the originate-to-designate model). This suggests, at systemic level, that consequently also the supply of credit provided by banks to the real economy is, at least in part, released by the ability of the same to collect deposits.

This only sums up what should by now be obvious to the reader. As the activity of wholesale funding cannot be carried out unless a network of actors is set up (i.e. broker-dealers, asset managers, and shadow banks), the shadow entities will not do this unless their aggregate profit will be enough to repay the risks. In addition, it should be noted that the involvement in the operation of a banking organization may open a route to the so-called "discount window" of central banks (and, in particular, to those of the FED in the event of a subject from the United States).

Now the analysis of the transactions that are characterized by their cross-board execution mode must be developed in more detail. The subject is substantially the operations in which some steps are carried out in the domestic market (usually those of origination, warehousing, and securitization of loans) and other international branch and in offshore centers (in particular those of the funding and maturity transformation of structured credit assets).

In analyzing these transactions, however, it is still convenient to classify them as external shadow banking subsystems. They are also closely connected with what is called the option for an international approach, which corresponds to an advantage in terms of specialization, given the opportunity to engage a wider audience of stakeholders (broker-dealers, non-bank specialist intermediaries, and private credit risk repositories). There is no doubt that the number of entities and the multitude of their operational experiences carry out a mix—of skills and expertise—able to formulate a varied offering of financial instruments (in terms of regulation and negotiation performance). This is, of course, in order to satisfy the global demand for CDOs.

It is fairly clear, therefore, that the internationalization of the shadow banking system is not always determined in relation to the research for regulatory arbitrage; as it could be the result of a process of specialization that casts some operators into "market niches." It will be safe to take the latter case as typical. There is, of course, no reason for supposing that this is not a case in which the organizers of an operation involve subjects who—not exercising a banking activity in its typical entrepreneurial forms—are more flexible and, thus, are able to quickly follow market trends (in function of a greater organizational efficiency, scale or objective economies, or other advantages arising out of the aforementioned specialization).[38]

Recent trends in the subsectors show a unique conceptual framework. The FSB perceived the existence of the aforementioned problems and is now trying to push its analysis to the point of solving it. It is clear that the relevant authorities cast the net wide, "looking at all non-bank credit intermediation to ensure that data gathering and surveillance cover all areas where shadow banking-related risks to the financial system might potentially arise."[39] This is why the analysis here shall narrow the focus to the subset of non-bank credit intermediation, where there are achievements able to justify both the increasing of systemic risk and regulatory arbitrage that moves capital from one state to another.[40]

2.5 Is this economic freedom or escape from regulation?

A different question is that of compatibility—or not—of the shadow banking system with the modern democratic system and, therefore, with the devices introduced for the protection of the general interest and the social equity. The content of private contracts at the base of the shadow operations and the rules of corporate governance under which the shadow entities are organized are, therefore, of particular importance. It might be more accurate, perhaps, to say that both of them (content and rules) affect the regularity of the shadow credit intermediation process.

The problematic definition of freedom-based banking business, contented between "entrepreneurial autonomy" and "public supervision," therefore, is evident, and not new.[41] However, with regard for the system under consideration, this question arises in new terms, in order to understand whether the access to this parallel system is (only) the exercise of their freedom and not (also) a misconduct of regulatory arbitrage.[42]

We can sum up what has been said here in the proposition that the shadows offer the possibility, to anyone who is interested, to break down the essential elements of banking transactions in single market operations, first, and, second, to rejoin them in order to implement a "more complex, wholesale-funded, securitization-based lending process."[43]

The economic and practical purpose at the basis of this possibility means that the exclusion of the exercise of freedom (to choose the sources of one's own funding) does not result in an unsustainable form of "resource grabbing" (actuated to the detriment of the local

Table 2.2 Economic perspectives

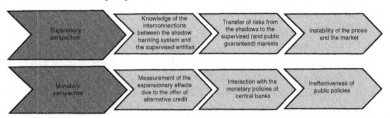

community). Similarly, that also has to be said of the organizers of operations, which should not interpret market activity as the possibility of exercising banking in the absence of the required authorizations (and, therefore, outside the rules; see Table 2.2).

To confirm this, it is important to note the significant frequency of speculative operations, which occurred prior to the involvement of systematically important banks (and other SIFIs).[44] There is a mixture between collection of savings and investment (of the same) in securities of the shadow banking system. The trouble arises, therefore, because of the multinational nature of the operators that facilitate the placement of these transactions in a cross border context (going beyond the scope of national banking supervision). It is most difficult to avoid grabbing and speculations.[45] It comes—one more time—from the lack of an international sovereign power able to control these operations (as a whole).

It, therefore, becomes possible to formulate an initial conclusion regarding the negative consequences that result from the delay in the establishment of a global financial governance and the difficulties that prevent jurisdictions to overcome national boundaries (in the regulation of the global cross-border economic and financial phenomena). These are difficulties and delays that prohibit the marking of a clear line of demarcation between freedom (of initiative) and anarchy (economic), despite the fact that—as we have seen—there are market failures requiring appropriate corrections.

In light of what has been said, the ways in which the interconnection of the shadow with the global financial system produces effects are clear, which end up interacting with the other key components of the real economy. This conclusion is not avoided by arguing that they point out the specific links between the shadow operations and the activities of banks (and, therefore, the mutual obligations arising from contracts that allow the assumption of counterparty risks bearing upon supervised entities that entered into business with shadow entities). Every assessment regarding the danger that the "collateral system" will be capable of transferring risks (of various kinds) to the banking system shall take into account these links, with obvious repercussions on the protection of savings entrusted to it.

Certain deeper perplexities, which may arise from the aforementioned conclusion, must be considered regarding the possibility that the areas of freedom of the shadow will end up hampering the orderly conduct of traditional banking and, more generally, the action of crime control put in place by national supervisory authorities. I am not referring, obviously, only to the obligation to ensure "equal legal settings" to all operators entering the market, but most of all to the necessary persecution of illegal activities that are running within the shadow banking system.[46] Moreover, there are all sorts of reasons why market-based financing will create a link between ethics (of finance), legality (operational), and value (economic). This is in order to ensure that it is not only the professional nature of the industry to avoid opportunistic behaviors (which go beyond the rules of the market), but also speculative actions to penalize the real economy (and, therefore, to reduce the overall levels of wealth).

2.6 The global nature and the riskiness of the phenomenon

So far it has been assumed that the affirmation of the shadow banking system is a result of specific economic reasons related to the convenience of the transactions realized in it. Consider carefully, however, that the growth of the phenomenon is also linked to the positive results achieved by the globalization of financial industry and, in particular, by the fact that the latter has promoted the cross-border provision of services through a network of relationships that allow free circulation of capitals.[47]

Moreover, specific importance must be attributed to the rejection of the "principle of execution in the stock exchange" of any trading of financial instruments. The stock exchange monopoly corresponds to the growth of a global market that creates competition between them and the trading venues: both in terms of price and quality (and, therefore, safety). In doing so, a context follows in which it is possible to apply new techniques, sometimes the source of a new kind of risk that is capable of developing speculative waves (that have fueled bubbles or cracks) and extending their effects beyond the individual economies of the firms that produced these operations.

Before addressing the profiles of the shadow banking system that interact with the globalization of financial markets, it is necessary to recall the thoughts of those who have argued that the liberalization of capital movements and financial innovation correspond to the formation of a new global financial system, in which the available resources circulate.[48] Adhering to this approach, there are many doubts as to the compliance of such a system with the founding principles of the contemporary legal systems. Consider that, in this case, there is not a sovereign power in place to ensure protection of individual rights and the common good. It is not clear how much longer this state of affairs can last, given the dangerousness of a system where there are no rules—simple and proportional—to oversee the stability of the mechanisms of capital circulation and the preservation of their social utility.

It should, in fact, be kept in mind that, at the G20 Summit held in Seoul in November 2010, the leaders of the industrialized countries have stated that the lack of controls on the shadow banking system is one of the open questions in the field of regulation of the financial system, highlighting how the strengthening of supervision of the same should be a priority for the FSB and other international standard-setting bodies.

This is a question for practical generalization rather than for pure theory. If there is some tendency to avoid *strict* rules, there may well be some sort of "rough relationship" between the recovery and *new* devices, incentives, and enforcement mechanisms. The latter appear necessary to constrain the operation of the shadow banking system with the objective of growth of the real economy, avoiding the risk of self-referentiality of the system. Therefore, the global nature of the phenomenon not only suggests the need to respond to the challenges posed by the crisis, but also the need to develop a model of regulation that can ensure equity to all the individuals who wish to finance their initiatives by direct recourse to the capital market.

It would be inappropriate, in forming any expectations (on the implementation of a supervisory mechanism on shadow banking), to attach a great weight to matters that are very uncertain. It is reasonable, therefore, to be guided by the disciplinary trend that—under the provisions of Directive 2004/39/EC (and, recently, of Directive 2014/65/EU)—qualifies the integration of EU financial markets. This is in relation to their instrumentality in respect of the efficient use of capital investment and, therefore, the presence of a functional link between them and the competitiveness of the real economy. Specific attention is, therefore, paid to the possibility that the activities carried out outside of public scrutiny affects the safety and stability of the markets in which savings transit (defeating the purpose of the disciplinary safeguards designed to ensure that the finance will achieve the objectives of social utility and allocative efficiency).

These are the facts about which one can feel somewhat confident, even if they may be less relevant to the issue than the fact that this does not solve the problem of the lack of regulation (in safeguard of the market failures) and of transparency (of the information) that qualifies the shadow banking system. Basically, the asymmetry remains between the socio-economic integration (financial circuits) and the political-institutional (of supervisors)—asymmetry that determines the coexistence of legal systems that are characterized by a different intensity of reaction to speculation and other degenerative forms of progressed capitalism.

In particular, there are no protections that should prevent *escapes* (of certain operators) from the rules governing the financial relations—including brokers, shareholders, management, and other stakeholders—in view of the general objective of "creating the value" in respect of the interests of individual counterparties and the overall balance of the economic and productive system. Hence, there is the risk that the activation of the channels of the shadow banking system is aimed at "opportunistic spill-overs" by the controls provided by the relevant regulatory framework (to evade certain limitations to financial transactions). It shall be considered here that—in such a case—it would give the course an asymmetry that cannot find any justification in terms of (legal) fairness and (economical) efficiency.[49]

On a more general level, it should be pointed out that the use of the aforementioned channels cannot allow operators to circumvent restrictions to the bilateral contract (or "synallagmatic" contract, in civil law systems), as it is possible to notice, for example, in the case of interest related to the laws against the usury. This reveals, therefore, the dangers associated with the orderly functioning of the shadow banking system, in relation to the fact that illegal forms of speculation would force the shadow process to achieve illicit financial transactions.[50]

This problem is amplified by the absence of a global legal system and by the lack of measures able to combat crime within the shadows markets. Conversely, if the latter are subject to the supervision of the Financial Action Task Force (on Money Laundering) (FATF), within the OECD, then there should be new forms of coordination between the states in given levels (albeit minimal) of legality. General action to tackle the international illegality should be added to the OECD's institutional tasks, also taking binding decisions and entering into agreements with its member states, non-member states, and international organizations (art 5, of the Convention of Paris, December 14, 1960). This means that the OECD should be the institution that is able to propose the measures to prevent the activities of certain financial firms—that are looking for very high returns, easily unsustainable in the banking system—degenerating into illegal activities.[51]

There is, however, not much to be said about the riskiness of the phenomenon *a priori*. The conclusion must mainly depend upon the actual observation of the shadow industry, markets, activities, and operators. This is the reason why the following analysis will focus on the doubt that the system in question can be an incubator for financial risks arising from maturity transformation methods (and their impact on monetary policies implemented by public authorities).[52] For convenience of exposition we shall assume in the ensuing focus on any risk profile of the shadow banking system that the main supervisory mechanisms are provided by the legal forms of self-regulation that form the basis of the business, as well as the safeguards of a general nature that surround the perimeter of the regulated markets.[53]

2.7 New freedoms and their problematic nature

To return to the immediate subject, the operations underlying the shadow banking system—and, more generally, the transactions that give content to market-based financing—raise issues of a new type. This is related to the orientation of the industry towards overcoming the—territorial, governmental, and personal—boundaries to which the financial markets were subjected in the twentieth century (when the role of national states and their sovereign powers were predominant).

If the reader has questioned the appearance of new freedoms for those involved in this area of finance, he or she must be answered that, at the base of the technical forms under analysis, "intermediation logic" can be found that, being outside of the regulatory processes, gives space to innovative interpretations of capital circulation.[54] But it must also be added that it makes use of schemes designed to give certainty of results to the relationship between the (alternative) finance and the objective of creating value (at the base of the market).

The consequence is the affirmation of a system characterized by the activity of responsible investors, called to act in conscious modalities, whose choices give content to the demand for credit. This consideration goes with the need to purify the shadow banking system from any pollutant action that—not responding to the abovementioned logic—pursues speculative objectives (with individual profit and sometimes predatory). There is no doubt, in fact, that the shadow banking system cannot represent an area of absolute freedom (able, that is, to accommodate transactions that would be illegal if carried out within the regulated systems). Hence, there is a need to identify the operational differences between the phenomenon under observation and the other circuits that allow themselves to be the birthplace of illegal acts.[55]

To date, chiefly held in mind is the idea that liberalization started in the 1990s and it may have seemed to be tacitly, assuming that regulators were aiming at the affirmation of an unlimited capital market, where everything is possible (or rather allowed). That is, obviously, not the case. Thus, we must also take account of the general tendency to go beyond the (legal) limits and (territorial) borders, when they are considered to be an obstacle to the objective of maximization of social welfare (which should qualify the economic processes of contemporary capitalism).

These considerations should lie beyond the scope of applicable regulations and supervising systems: and they must be relegated to their right perspective. If I may be allowed, I assert that in the shadows a new kind of freedom manifests the possibility—for the organizers of a credit transaction—to change the legal and financial structures of the vehicles in order to reach their own goals.

As the organization of the shadow credit intermediation process can involve the freedom of using derivatives, the risk of predominance of speculation (and other unfair practices) does, however, increase. In other words, in the shadow banking system, the activities of loan and ABS warehousing can be executed also through synthetic transactions, with obvious implications in relation to the complexity of the phenomenon. Therefore, I cannot fail to mention that the effective exercise of this freedom requires new standards of transparency (at present difficult to find in financial statements and other disclosure documents) of the special purpose vehicles in use in international practice.

This is another way to reach the conclusion that these freedoms undermine beliefs, supervision powers, and intervention methods that were strengthened during the twentieth century. From a professional point of view, in overcoming the model of the commercial bank, the shadow entities are subtracted from the well-known (and experienced) authorization mechanisms, with the risk that businesses will not equip themselves with a permanent (or better, sound, and prudent) organization, subject to a statute and internal controls able to monitor and mitigate the operational risk (and the moral hazard propensities). In particular, as we shall see, the shadow banking entities:

- are brokers who are outside of the legal frameworks provided by the national legal systems;
- often operate in offshore centers (where they also realize the sale of assets required for the production process of financial instruments); and
- choose in autonomy the markets that are marketing the CDOs (in order to meet the demand of cash pools).

Wanting to draft another initial conclusion, it may be emphasized that, according to the affirmation of spaces not taken into account by the political authorities, the limits of contemporary democratic systems are recognized, anchored to the borders of national states (in the case of the European Union, confined to the regional level).

It is hardly necessary to mention that in the shadow banking system the only powers that find widespread application are of a private nature and are available to the parties. But the latter independently predispose the contractual settlement of the transaction or decide to incorporate their affairs in a given system (being able to choose—in the negotiations—the forum for the resolution of any dispute). Beyond any enforcement problems (deriving from the possible escape of a subject from the application of the aforementioned powers), it appears evident that such structure is not able to achieve conditions of equality, leaving the "strong contracting party" free to adapt the operation to their *desiderata* (beyond any constraint of fairness). It is likely, ultimately, that—in the shadows—the evanescence of authoritative controls can support the predominance of (economic-bargain) power on (personal) rights.

Even apart from the issues arising from the freedom of speculation, there is instability caused by the problematic nature of the forms in which the globalization process (and the associated risk proliferation) finds affirmation. It is imperative to take into consideration some recent analysis that highlights the relationship between the regulation of the conduct of those who intermediated capital (and, in particular, large international banks) and the purpose of ensuring high levels of safety in financial markets. It seems clear that the latter are conditioned by a high detrimental potential—on the socio-economic level—of certain market-based transactions, as well as by the difficulty to counteract any "domino effect" (or rather, contagion) or to prosecute any fraudulent conduct (dressed up, disguised as shadow operation).

Most of the ideas of promoting a new impulse to public intervention are aimed at creating transparency in the financial market and can be taken as a result of specific analysis of benefits, not as the outcome of a desire for "statism."[56] The goal of this intervention is identified—in terms of regulation—in moving supervision beyond the sanctioning (attemptable in case of injury of others' legal positions) and then towards the objective of ensuring the overall stability of the channels through which the capitals circulate (and, therefore, also those of the shadow banking system).

3
Shadow Banking Entities

By shadow banking entities, I refer to the subjects operating in the process that transforms "loans to real economy" into "financial instruments" (see Table 3.1).

Hence, this chapter, firstly, addresses the questions regarding the form and the legal status of the special purpose vehicles and the entities based on collective investment schemes (focusing on "money market funds" or MMFs).

At the same time, I clarify that the notion of a "shadow bank" refers to an economic idea, and not to a legal entity.

3.1 Special purpose vehicles

The "shadow credit intermediation process" is based on the presence of one or more special purpose vehicles (SPVs), in the legal form of companies, designed for the acquisition of assets (i.e. the credits arising from the loans) and for the offering of financial instruments (i.e. ABCP, ABS, and CDO).

I can best introduce what needs to be said on these SPVs by highlighting that the arrangers of the shadow banking operations establish such vehicles in order to account (among the assets and the liabilities of a regular company) for both the credit rights underlying the issuing, and the debt rights arising from the aforementioned offerings.

As has been shown in the previous chapters, the basis for the lawfulness of this process is very precarious, because of the lack of public supervision within this industry. According to these findings, the investigation—from a regulatory perspective—must focus on the role of the shadow banking entities, because they influence the economic determinants of the shadow banking system and the relevant flows of capital.

Table 3.1 Features of shadow banking entities

Special purpose vehicle	Shadow bank	Captive financial company	Conduit	Structured investment vehicle	Shadow fund
• The SPVs, on the one hand, proceed in granting the loan and, on the other, issue the ABCP, ABS, and CDOs. • EU Regulation no. 1075/2013	• SB is a business (and not a legal entity) in charge of the "lending activity" that takes place (i) outside of the areas of public supervision, and (ii) in the absence of direct relations with the monetary authorities	• It is a financial intermediary that exercises the activity of granting loans to facilitate the purchase of goods or services offered by the entrepreneurial group to which it belongs	• Single-seller and multi-seller entities purchase asset-backed securities from a single or from more than one originator: these entities realize a stage useful to the shadow credit intermediation process	• Corporate body used because it can remain off balance sheet and not be consolidated by bank • Commission Decision of 4 June 2008 on State aid C 9/08	• The arrangers use investment funds in order to complete certain stages of the shadow credit intermediation process • EU Directive 2009/65/EC (so called UCITS Directive) • EU Directive 2011/61/EU (so called AIFMD)

This brings me to the current topic. First, consideration should be given to the equilibrium of demand and supply, and then to the equilibrium of the intermediary activity carried out by non-traditional operators. In brief, this means that the intervention of an SPV, on the one hand, assists in granting the loan and, on the other hand, helps to issue the CDOs.[1]

Thus, from a regulatory perspective, the vehicles used to start the securitization will be the "initial creditor" (of the debtors who signed the loan contracts, originated or received through a credit assignment agreement), while the (other) vehicles used to proceed in the offering of the CDOs will be the "final debtor" (towards the investors, the market, and the public supervision authorities).

It is obvious that between these two vehicles (placed at the beginning and the end of the process) there shall be other SPVs, which buy financial instruments to perform operations of resecuritization (in order to achieve the effects of transformation that characterize the shadow banking system).[2] In this case, even if these SPVs can have only private relationships (in a business to business loop), they cannot be considered out of the scope of public supervision, because of their influence on the performance of the shadow process.

It may be convenient, at this point, to say a word about the role played by any of these SPVs within the shadow banking system. Practically, we must identify both the *burden* of proper implementation of any step by any SPV and the *interest* in developing the most suitable asset configuration to support the offering of CDOs in a way that the investors are willing (and able) to purchase, according to the market demand. Hence, there is the need to understand the constraints to prevent conflicts of interest, and to ensure the correct development of the operations.[3] Therefore, in the shadow banking system, the presence of such a vehicle is required for the organization of its capital intermediation process (and then for lending credit to the real economy or raising funds on the capital market).[4]

From this perspective, the research must linger on the activity and organization of the SPVs, in order to verify the compliance (with the current financial regulation) of the legal design of these vehicles (being that they are the "center" to which legal relations have to be allocated for the realization of one of the stages of credit, maturity, and liquidity transformations provided by the shadow credit intermediation process).

Considering the subjective profiles of the SPV, it should be clear that its legal structure shall be designed to support only one type of business: the one that gives content to a single stage of the shadow process. This means that any part of the operation can be referred to as the economic paradigm of "subcontracting," being functional to the production of a semi-finished financial product (i.e. ABCP and ABS, mainly), to be structured according to the characteristics specified in the intermediation project developed by those who organize the abovementioned process. Hence, the SPV is committed to its own responsibility and, therefore, shall follow the principles that characterize the contractual law applicable to the relationships that are established with third parties.

Moreover, if the SPV is owned by a credit institution or is part of a banking group, the assets and liabilities of the vehicle can be consolidated in the financial statements of the relevant bank or holding (and, therefore, will be subject to the rules of prudential supervision and capital adequacy, as well as the traditional monetary backstops).[5] In this case, a bank-owned SPV cannot be included in the shadow banking system, because of the comprehensive nature of the banking supervision. Therefore, any further consideration of the possibility for a bank to establish and participate in a shadow banking entity will be meaningless and lead to error.

Furthermore, when a shadow operation is promoted and managed directly by a bank, there is no reason why the latter shall not be involved (and, therefore, are called for action) in case of default. And this consequence does not only imply its reputation, but also the legal effects provided by contractual relationships or other connections (also synthetic, as in the case of credit default swaps). Obviously, the abovementioned consequence will not be produced by SPVs that do realize a sort of outsourcing, and then fall outside of the economics of any bank (even if in the offering of CDOs).[6]

It is reasonable, further, to refuse the elusive practice that leads banks to use vehicles designed to conceal any non-performing asset. This practice, in fact, is an attempt to circumvent the prudential rules set by the Basel Accords and, therefore, adversely affects the stability of the banks (which remain in the position of "parent bank").[7] Nothing allows any bank to hide in the shadows the responsibilities that, in various national legal systems, arise to unite all the subjective proliferations that maintain connections (relevant for prudential purposes) with the credit institution that has originated the relevant obligations.[8]

Hence, here the focus will be on the corporate governance of the SPVs, as the correct performance of their business functions (of management and control) is necessary for the effective achievement of the credit-transforming effects of the shadow process. This performance, in fact, has externalities both in terms of transparency and convenience of market rates. According to the relationship that exists between "good standards of corporate governance" and the "quality of the asset-management of financial firms,"[9] the following analysis shall conclude by suggesting the monitoring of the internal rules of the vehicles used to set up the shadow banking operations.

We must, then, take into account the impact of the new regulatory wave, based on the shift from a one-size-fits-all approach to another focused on the fluidity of governance trends.[10] In this context, the organizational layouts of SPVs, in addition to satisfying the interests of those arranging the operation (and, therefore, providing the initial capital injection), must ensure the safe and sound management of the assets and, consequently, the sustainability of the projects made for the offering of the financial instruments (even if the latter circulate only in the intermediate stages).[11]

This leads to the consideration of whether the choice, carried out by the FSB, to promote a strengthening of the supervision on the governance of any shadow banking entity is fully justified; this choice clarifies the attempt to prevent any internal problem (of a single vehicle) affecting more than one operation and, therefore, the whole dynamics of shadow intermediation. Consider that the FSB, in promoting the predicted strengthening, will be able to rely on the basic principles of the most recent supervisory provisions for banks and, in particular, those contained in the Directive no. 2013/36/EU.[12]

If this possibility is considered carefully, there can be no doubt that this must be the end of the market praxis of *snubbing* the governance structure of the SPVs. There will be the opportunity to satisfy the need for new sound practices.

On a subjective level, the presence of SPVs in the shadow banking system requires the implementation of governance arrangements capable of ensuring efficient risk management functions. This does not mean that the supervisory authorities should restrict private autonomy in the definition of such devices, but that the regulators should adopt safeguards able to ensure effective supervision on the part of the organization in charge of asset management (according to a criterion of proportionality), in line with the current risk culture, which should preside over all the actors in the capital market.

Of particular interest is the Italian experience that has directly addressed the issue of securitization and other similar transactions (in line with the directions of the CRR and, more generally, with European law).[13] The Italian Supervisory Provisions—contained in the Circular no. 285 of December 17, 2013 of the Bank of Italy—regulate any (net economic) interest of the bank in a single transaction of this kind, as well as the positions taken in respect of securitization (and, therefore, the resulting impact on the assets and liabilities of the credit institution).

We must take into account that this Italian set of rules creates a duty of "adequate verification" (per exposure) of the issuer, asset class, and level of concentration. This duty, in fact, can promote the proper allocation (of risks and opportunities) in the SPVs' balance sheets and, therefore, the production of reliable financial reports (such as to provide information to the market).

Unfortunately, certain European national supervisory systems perceived the existence of such a problem without being able to push their intervention beyond the regulation of individual securitizations (and without providing a more general plan that can encompass the entire production cycle of the CDOs).

It is, then, understandable that the strengthening of the supervision on the SPVs must take into account the current configuration of national supervisory systems. Hence, consideration must be given to the fact that the current European regulation limits the duty to report the analysis of (credit) risk to all of the positions taken by the bank to the SPVs involved in the securitization. This is a choice that fails to capture the actual risk profile of the mentioned entities; hence, the need for a rule introducing an additional weighting factor (in art. 407 CRR) compared to the dangers that characterize the phenomenon under observation.[14]

It is convenient to mention, at this point in the analysis, that the model at the base of the European supervision seems unable to ensure an effective supervision of the SPVs that operate in the shadow banking system. I refer, in particular, to the limited effect of the discipline contained in Regulation (EC) no. 24/2009 (adopted by ECB), concerning the recognition of the assets and liabilities of financial vehicle corporations engaged in securitization transactions. The European system remains, undoubtedly, within the limits of a monitoring system that does not innovate the pre-existing structure (laid down by Regulation (EC) no. 2533/98, concerning the collection of statistical information by the ECB).

Even if this monitoring has positive externalities, we can understand that the aforementioned rules reflect the idea that policies of restrictions can be "treacherous instruments," since administrative incompetence and other intrinsic difficulties (in regulating the banking activities) may divert by giving results conflicting with those expected.[15] These rules can be considered, then, as the result of strong presumptions against regulatory restrictions to financial markets.

Conversely, as explained in the previous chapters, it is important to try to justify the implementation of new regulatory systems with the possibility of reaching gains in efficiency. Consider, then, that it is necessary to rely on an exhaustive regulatory framework; one that is able to ensure the effectiveness of the governance of any SPVs (in order to have a safer process of financialization and re-financialization). Otherwise, the vehicles can not be considered as an active subject of the shadow process, but a mere legal fiction designed to separate and fragment the responsibilities underlying the shadow credit intermediation process (and, as will be shown, to limit the liability of the arrangers of the shadow banking operations).

3.2 Shadow banks

The words "shadow bank" suggest that, in the system under observation, there is a subject performing fund raising and lending by acting (i) outside the areas of public supervision, and (ii) in the absence of direct relations with the monetary authorities.[16] According to a mere textual approach, it shall be expected that a company will take up the business of a credit institution without having autorization or any relation with its central bank. But, according to the most common regulatory frameworks, this is unlawful.[17] On the contrary, in the shadow banking system, more companies are observed dealing together to lend money, create securities, and collect capitals on the financial markets.

Currently we are in a situation of *binomial formula* that summarizes the economical function of the shadow credit intermediation process, unifying in a single "black box" a reality that, in legal terms, involves different subjects and scope in the process that goes from the organization of any shadow banking operation to the satisfaction of (credit) demand and supply (of CDO securities).[18]

However, the idea behind shadow banking is sound. Only in appearance can it be considered as an oxymoron, since it merely highlights the alternative relationship between traditional banks and the entities that realize the form of market-based financing under observation.

It brings, therefore, into consideration the *nature* of that expression, which—beyond the terms used—refers to the set of subjects who, instead of leaving free space to market mechanisms, coordinate "a wide range of securitization and secured funding techniques such as asset-backed commercial paper (ABCP), asset-backed securities (ABS), collateralized debt obligations (CDOs) and repurchase agreements (repos)."[19]

Let me state in my own terms what now seems to be the agreeable elements in the notion of shadow bank. I will then compare this with the actual assumption of both global regulators and other international supervisory bodies.

It should be clear that, within the system under consideration, there is not a "business entity" that can be identified as a company that intermediates funds, but a set of companies who work together in order to execute the shadow credit intermediation process. It should be noted, then, that the term *shadow bank* can be used only as a single point of reference, useful in the economic analysis to compare the market-based financing with the traditional banks; and this, both to evaluate the volumes and to estimate the risks. Conversely, in a judicial perspective, we cannot refer to a shadow bank, but to *shadow banking entities*, which should be considered as autonomous subjects and, as such, can operate in any given financial market.[20]

However, the concept of shadow bank can be helpful in any accurate analysis of current regulation, as it can help in measuring the impact of regulatory standards on the methods of intermediation in use and, in particular, on the interconnections between these entities (as a whole) and the global financial system. Any assessment regarding the aptitude to *select* the creditworthiness and to *evaluate* financial instruments able to circulate regularly can be referred to this concept.[21] Therefore, this concept must be used while taking into account the fact that, from a legal perspective, it indicates a "flow of operations" rather than a legal entity (which carries out an intermediary activity).

In other words, certain economic analyses assume as a hypothesis the unity of a case that, on the legal and social level, shows itself as a collective (though fragmented among subjects who are exposed to different risks, and sometimes belong to different national legal systems).

In addition, consider that the macroeconomic model developed by the Bank of Italy considers the shadow bank not as a banking firm (which internalizes the process of credit intermediation), nor as an operator who, on its own, allows the circulation of capital according to market mechanisms. It is a logical driver that leads to the unity of the set of subjects that "hold loan pools comprised of primary security bundles acquired from many originating commercial banks (other than the banks owned by their home household), financed by a combination of inside equity and ABS." This is based on the assumption that "securitized assets are held within the financial system, rather than being distributed to unlevered investors (households, in our model)."[22]

Such a model helps to understand the need for a regulatory intervention that goes beyond the monitoring of the "basic real business cycle model" and the single shadow banking entities, in order to identify the data able to verify whether "the shadow banking system is economically valuable because, by transforming illiquid loans into tradeable assets, securitization allows collateral to be used more efficiently."[23]

To summarize: in the shadow banking system subjects who, by internalizing the lending activities, reflect the organization and activities of traditional banks cannot be identified. This cannot be lawful, hence, companies exercising banking activity in the absence of the required authorizations are subject to the sanctions provided by the traditional legal system (notable among which is the expulsion from the capital market).[24] In the light of the foregoing, it is easy to understand what the contribution of this concept has been to defining the strengthening path of the vigilance.

From another perspective, the abovementioned focus on the concept of the shadow bank can explain the reason why I will not refer to shadow banks as "a kind of credit intermediary that lies outside the range of much banking regulation." One is aware, in fact, that—from a legal perspective—several shadow entities are involved in the process that "has created a kind of money that is likewise beyond reach of central bankers' traditional instruments of oversight and control."[25]

The *helplessness* of the concept of the shadow bank to build a system of supervision responsible for the monitoring of capital flows and for the management of the risks of the shadow banking system can be deduced from this. Hence, the profiles of responsibility and the risks of any shadow businesses must be evaluated with specific regard to the legal nature of each shadow banking entity and to the type of activities exercised. In this case, then, no reason has been found to suggest the *usefulness* of an extension—constant and uncritical—of the mechanisms used in the supervision of banks.

3.3 Other shadow banking entities

The presence of many companies involved in the transformation process of loans into CDOs is outstanding evidence of the shadow banking system. These companies cooperate with the SPVs and only as a whole can they be considered as a shadow bank.

If, in the economic analysis, these companies can be treated as a unity (i.e. the aforementioned black box that transforms inputs into outputs), this simplification shall be refused, because it prevents the understanding of both the individual responsibilities and the possible conflicts of interest that arise during the shadow process. It is, therefore, necessary to dwell on the role that each company plays, in order to understand its actual business and to verify if it is already subject to a form of supervision (according to the applicable traditional regulation).[26]

Although this state of affairs would be quite compatible with some measure of safe and sound management, the subprime mortgage crisis has already highlighted the importance of the phases of loan selection, in which the operator that creates the first SPV lays the foundations to prevent a massive default of a set of debtors from resulting in a general reduction of the securities' market values and, therefore, in a "crisis of confidence" that can lead to a *escape* of final investors (with obvious negative effects on the overall functioning of the system under consideration).[27] I feel confident that this operator is mainly responsible for the evaluation of creditworthiness and for the implementation of checks and balances that are necessary to understand the quality and, therefore, the reliability of the "raw material" (i.e. the loans) placed into the production cycle of the CDOs.[28]

Hence, there is the need to devise a mechanism of self-control to protect, not only the operational correctness, but also the smooth running of the shadow banking system, in order to avoid opportunistic behaviors pursuing an undue increase in credit volumes (aimed to increase the revenues of these financial firms). This mechanism shall align the individual interest of profit in order to promote the production (and the disposal) of safe and reliable credit rights (to avoid, from an early stage, elements of instability leading to a market failure).

Special attention must be focused on the figure of the "captive financial company," which exercises the activity of granting loans to facilitate the purchase of goods or services offered by the entrepreneurial group to which it belongs.[29] The conflict of interests afflicting these kind of companies (to be considered in relation to the business objectives of the parent company) is evident, to which is related the risk that the valuation of the creditworthiness suffers the negative influence of the desire to increase business volumes, revenues, and market shares (of the group). Even in this case, therefore, control mechanisms (internal or external) of the qualitative assessments in view of granting credit become necessary.

From another perspective, we must consider the "single-seller" and "multi-seller" businesses, meaning those involving subjects who issue the securities to be offered to the market through the transformation of one or more "classes of receivables" by one or more originators (which were part of a sale of the aforementioned assets).[30] It goes without saying that these entities may use one or more SPVs (newly created) in order to segment their operational risk, preventing the default of one operation affecting all their assets (or at least the regular course of the other operations referable to the same seller). Also, in this case, we understand the usefulness of the SPV, favored in the praxis to other mechanisms of legal separation of assets.

Obviously, any single or multi seller can operate as a free and independent entity (by performing only one phase of the CDO production cycle and providing intermediate goods and services) or act under the direction and coordination of a third party, responsible for the whole organization of the relevant shadow banking operation. In both cases, the sellers will have to constitute stable relationships—whether in contractual or participative terms—with other parties involved in the production cycle, in order to ensure the liquidity and the credit enhancements necessary for the reliability of the shadow process (and, therefore, receive an appropriate rating assessment for the financial instruments issued). The same has to be said for the operations based on "conduits," in which the exclusivity of the social object impacts on the pricing of financial instruments and, consequently, on the market rates trends.[31]

Therefore, this brings into consideration the problem of economic efficiency of a process that provides the intervention of more entities, wherever it appears necessary to verify the effectiveness (and fairness) of a production cycle that fragments any activity and, therefore, the relevant responsibilities (including an indefinite number of entities whose assets are unlikely to be attacked by actions taken by the final purchasers of the securities that are placed on the market).

Needed in this context, therefore, are both a rule of judgment to address any individual responsibility following an opportunistic behavior (forcing the leverage or one of the other parameters at the basis of the shadow banking system) and a special regulation to ensure the transparency of transactions by improving the traceability of the raw materials (i.e. the loans) entered into the production cycle of CDOs.

Other relevant participants to the shadow credit intermediation process are named "structured investment vehicles" (SIVs). That is, the ones in the form of an entity (even as a limited partnership or a corporation) with a scope reduced to pursue very specific purposes or to act for a short period of time, usually able to separate certain financial risks (mostly the risk of bankruptcy).[32] The legal structure of these entities allows their shareholders to exclude them from the scope of consolidation (of the institution that promoted their establishment) and, therefore, they are neutral for the purposes of the balance sheet of a supervised banking group.

It should be kept in mind that, in the banking industry, SIVs are often associated with the practices of circumvention prudential supervision. Through their constitution, certain banks try to separate specific assets in order to reduce the capital requirements set by the relevant regulations (taking advantage of the existing asymmetries between the weighting of the loans and of the ABS or CDOs). Therefore, the practices through which "the banks are able to fund loans at lower rates than those of their own" should be neglected (as they are related to the shares of liquidity provided by the rules of banking supervision).[33] Obviously, these practices cannot be considered as a safe part of the shadow credit intermediation process. As was explained at the beginning of this investigation, the subjects that lend themselves to the avoidance of special legislation (and, in particular, the rules on the basis of the Basel Agreements) cannot be included in the scope of this analysis because they do not take as a business purpose a lawful financial activity, but they do contribute to the implementation of choices contrary to the purposes of banking supervision.

In conclusion, it can be said that the concept of shadow banking entities identifies a numerous set of subjects characterized by the independence of the organization (managing and accounting), as well as the specificity of the activities carried out (in respect of the production cycle of CDOs). Hence, there is a need to dwell on the content of such activities, in order to understand what are the responsibilities (of the entities in question) resulting from participation in shadow banking operations. Consequentely, analysis will be carried out in Chapter 4 in order to verify if the path—for the strengthening of the supervision—proposed by the international bodies reaches the goal to control the shadow operations of the aforementioned entities.[34]

3.4 Shadow funds

I have mentioned in passing that the liberalization of financial markets might be more favorable to innovation than the previous regime has been. It is worth repeating and emphasizing this aspect with regard to the versatility of the entities that are active in the collective portfolio management industry, which allows the arrangers to call for the intervention of asset managers and alternative investment funds in order to complete certain stages of the shadow credit intermediation process. These funds (and, in particular, the MMFs), in fact, allow the design of "advanced organizational structures," resulting from the integration of a clear regulatory framework and a broad freedom accorded to forms of self-regulation (which give content to "fund regulation"). This is especially the case in Europe, following the adoption of Directives 2009/65/EC (so-called UCITS Directive) and 2011/61/EU (so-called AIFMD).

It is not in doubt that the specificity of this regulatory complex not only allows the intervention of professionals (who manage resources collected from third parties), but also the involvement of legal structures known to the financial market as the "undertakings for collective investment in transferable securities" (UCITs) and the alternative investment funds" (AIFs). Nevertheless, the shadow process takes advantage of the presence of the asset managers, whose organization is capable of ensuring high levels of efficiency in the financialization of the assets held by the managed funds (i.e. loans, ABS, and ABCP). In this context, the asset manager can provide the arrangers with the capacity to issue financial instruments able to circulate on the regulated market (i.e. mutual funds shares, to be considered as regulated securities).

In principle, an asset manager can set up a fund to invest directly in credit (arising from the loans) or in securities backed by credit (issued by an SPV). It is, therefore, necessary to deepen the research on the role of the asset managers within the shadow banking system. It goes without saying that, with regard to the transactions involving the funds, both the organizational structure of the managers and the fund regulations shall be taken into account. In this context, the program of activities (set by the asset manager) shall be preordained to ensure a professional management of the assets held, and then a transparent relation with the investors.

In any case, taken into account here is the fact that the fund regulation represents a manifestation of the arrangers' will, given that the investors can perform only self-control tasks.[35] Hence, the importance of the management company and of the AIFM being considered as responsible for the stage in which a fund operates, even in the function of an express provision of law contained in art. 6, Directive 2009/65/EC and art. 6, Directive 2011/61/EU.

With regard, then, for the possibility of finding a link between the funds and the monetary policies (in the euro zone), it is necessary to ascertain whether the former can be included among the "eligible counterparties," identified by the *Guideline of the European Central Bank* of September 20, 2011 on monetary policy instruments and procedures of the eurosystem, where it is stated that "for outright transactions, no restrictions are placed a priori on the range of counterparties."[36]

It should be kept in mind that, in the generality of the eurosystem monetary policy operations, "only institutions subject to the Eurosystem's minimum reserve system are eligible to be counterparties," as stated in the provisions of art. 19 of the Statute of the ESCB. This is the reason why Regulation (EU) no. 1358/2011 of the European Central Bank of December 14, 2011 (which provides rules on the application of minimum reserves) circumscribes this possibility to credit institutions.

That said, if, on the one hand, asset managers and funds are subject to financial supervision, on the other hand, the same cannot access the generality of monetary transactions (except for certain outright transactions of the ECB). It is difficult, therefore, to place the funds in the "outer perimeter" of the supervision or in the shadow banking system, where the definitions adopted by international authorities do not offer clear guidelines. In any case, there is no doubt that funds are subject to the control of financial authorities, and that they cannot be traced *tout court* to the type of shadow banking entities (and, therefore, placed in a context external to the perimeter of the public intervention). This, however, does not exclude the need for a strengthening of the supervision in the field, such as to complete the action started by the mentioned Directive 2011/61/EU in order to prevent the participation of a fund to a shadow banking operation affecting the nature (and, consequently, the risk profiles) of the organism in question.

3.5 The particular role of money market funds

It is important to highlight that the FED and the European Commission focused on money market funds (MMFs) when they analyzed the shadow banking system.

These funds have been defined—by the US Securities and Exchange Commission—as "a type of mutual fund that is required by law to invest in low-risk securities." For this purpose its is specified that investments in this type of instrument "have relatively low risks compared to other mutual funds and pay dividends that generally reflect short-term interest rates," understanding that "unlike a 'money market deposit account' at a bank, money market funds are not federally insured." Nevertheless, in the United States, these funds are regulated entities "primarily under the Investment Company Act of 1940 and the rules adopted under that Act, particularly Rule 2a-7 under the Act".[37]

In Europe, the definition adopted by the European institutions to identify the money market instruments, according to Directive 2009/65/EC, refers to "instruments normally dealt in on the money market which are liquid and have a value which can be accurately determined at any time" (art. 2). However, it should be noted that the proposal for a European framework designed for MMFs is still being discussed. Therefore, steps towards the adoption of a definition similar to the one proposed by the Commission is expected; according to which "Money Market Fund (MMF) is a mutual fund that invests in short-term debt such as money market instruments issued by banks, governments or corporations."[38]

Therefore, a regulatory framework can be identified that has not yet found a complete configuration. It is difficult to foresee if EU institutions will be able to refine their organization in order to supervise the MMFs (given that they are exposed to the contagion of the risks that originate in the shadow banking system). What can now clearly be seen is that, currently, there are no rules able to protect MMFs' investors when involved in the production cycle of CDOs. This was noted by the FED's identifying of the MMFs as the main recipients of the supply of such securities to the extent of "about $1.6 trillion, much of this from money market funds and securities lenders, through tri-party repos, leaving aside additional funds sourced from asset managers and other investors through other channels."[39]

A similar conclusion is reached also by the European Commission, which aims at increasing controls on the quality of the assets held by these MMFs and, therefore, on the reliability of assets that they purchase in the shadow banking system.[40]

It is clear how the analyses just mentioned highlight the (quantitative) impact of the investment made by the MMFs in the shadow banking system, which—in conformity with applicable legislation—are guided by the decisions of asset managers (by appling the predetermined policy set by the fund regulation, and then investing resources collected from third parties through the issuance of the fund's shares). And this is true even if, in this case, the investors are "mostly corporate treasurers who need to hold large amounts of cash on a short-term basis and who do not want to put all of their cash in one single bank deposit account."[41]

There is also no doubt that these managers pursue an activity that is subject to financial supervision. They monitor the compliance of investments to the "fund regulation" (which is also approved by the competent supervisory authority and signed by the participants), and then they provide the main safeguard for the protection of the confidence in financial markets.

The need to protect the shadow banking system from the replay of runs (similar to the ones experienced by MMFs) goes without saying. Hence, it is expected that the regulator will realign the incentives that qualify this industry and, therefore, will correct the previous, fallible market structure. Therefore, an additional way of intervention can be identified, that is anchored at "minimum liquid asset requirements... [and]... temporary suspensions of withdrawals and redemptions in kind" and designed to reduce the pro-cyclical trends of the supply of credit by the shadow banking system.[42]

That said, it is good to have regard also for the directions provided by the Bank of England, which suggest that "Europe should want a global standard or at least a globally consistent approach to money funds... [and]... authorities in Europe and elsewhere will need to think through what if any measures we could sensibly take to make our part of the global financial system more resilient to the fault-line that the money fund industry currently represents."[43]

It is also useful take into account that this British proposal goes beyond the redefinition of the scope of the capital adequacy parameters required by the global regulatory framework for more resilient banks and banking systems (and, in particular, of the funding ratio). It is projected toward the new levels of *prudence*, summarized in the objective of "taking shadow banking out of the shadows to create sustainable market-based finance."[44]

Accordingly, the expectation is that the regulatory policy will move towards the redefinition of the criteria used to set the relevant parameters (and, in particular, of the net asset value), and then to place new organizational safeguards in the investment and risk management policies (also in order to avoid pro-cyclical feedback overly sensitive to certain market parameters). This will avoid the problem of certain stretchings affecting the quality and efficiency of markets that have achieved high levels of development. It will also benefit both the industry of collective portfolio management and the shadow banking system.

4
Shadow Business of Banks, Insurance Companies, and Pension Funds

In this chapter, the focus is on the role played by the commercial and merchant banks in the shadows, and I identify the need for accurate internal controls designed to avoid excessive risk taking and moral hazards.

Firstly, the chapter looks at the several financial firms and credit institutions involved in the multiphasic process of the shadow banking system. Then, it takes into account the activities of insurance companies and pension funds, which can access this alternative system to provide guarantees or acquire the CDOs (being, in this way, exposed to the full shadow banking risks).

It should be obvious, at the end of this chapter, that the access to the *shadows* of these companies is lawful only if it is contained within the risk constraints provided for by prudential regulation. Consequently, the system of internal controls and the public supervision shall avoid hazardous operations aimed only at improving the profits.

4.1 Multiphasic shadow credit intermediation process and the roles of traditional operators

A financial entrepreneur intermediates capital (for a certain margin of interest) or provides services to its clients for a certain income, but he or she probably won't care if its *output* (i.e. loans or services) goes into the production cycle of the shadow credit intermediation process. Neither, will the entrepeneur take into account that his or her involvement in the shadow banking system helps this system to grow and, then, compete with the banking industry. The financial supervisors do not seem to be upset by considerations of this kind, and appear anxious to set up a framework of checks and balances.[1]

On the contrary, it is clear that the involvement of credit institutions, insurance companies, and other subjects of the financial industry requires regulation and control. Obviously, the public intervention shall take into account that this involvement is not simultaneous, continuous, or permanent, but it follows the course of the sequential stages according to which, once the initial loan is originated, operations of warehousing are alternated to securities issuance. And, as has already been said, this goes on until the final product (i.e. the CDOs) are offered in the market and placed with the investors.[2]

Bearing in mind that the ultimate goal of this process is the credit intermediation, the will of regulated firms (banks and insurance companies) can be understood to provide services to the shadow entities in order to increase their profits. Consequentely, the examination of the type of individuals that realize wholesale funding must take into account the "complementarity and reciprocity" of the interests of the (supervised or shadow) parties involved in the transactions in question.

Over time, the requests arising from the *shadows* have allowed the specialization of the operators and, therefore, a differentiation of the above competences, with the obvious consequence that—at present—it is possible to distinguish the role of the operators, and then differentiate them in relation to:

- the stage of the process in which these operate;
- the nature of the activity provided; and
- the types of risks faced.[3]

It is good to point out, then, that the shadow process does not only involve SPVs (or other entities who commonly perform transactions of securitization and re-securitization of bank assets). As aforementioned, it also involves other operators who are able to provide, on the one hand, the supply of loans (granted to subjects of the real economy) and, on the other hand, the demand for safe and liquid financial instruments (arising from the treasuries of multinational companies or other entities that hold cash sums).

The option to include in the shadow banking system all the operators who contribute to the realization of its process, regardless of the types of relationships (contractual or participatory) that give content to the phenomenon under examination, can be understood. Consequently, it would be detrimental—in the understanding of the phenomenon—to attempt to classify this specific case in the traditional categories of the "temporary consortium" or the (business or economic) group, which are poorly suited to discipline in this system (i.e. organized by a plurality of independent entities, both at economic and legal level).

This is the affirmative answer to the question of whether the subjective profiles of the shadow banking system can interest a law researcher more than they have been taken into account in the economic analysis. Hence, it appears to be clear, because, in the opinion of the European Commission, supervised intermediaries can be considered *partakers* of the shadow banking system.

Recall that shadow entities are—in principle—excluded from the direct access to the liquidity injected into the market by central banks (and, in particular, to the so-called FED's discount window or to the ECB's long-term refinancing operations), as well as other public resources or deposits guarantee systems (including the federal deposit insurance). This is in order to clarify that there are no rules allowing any subject of the shadow banking system to require *tout court* public intervention to cope with the negative effects due to the occurrence of the related liquidity risk. This reflects the option to play outside the rules and, therefore, in the absence of backup facilities that are able to socialize any losses occuring in the system under examination.[4] And this is, therefore, one of the main distinguishing features between the shadow banking entities and the traditional operators involved in the shadow process.

The outstanding financial crisis in which we (still) live shows the obvious limitations of such an approach. The exclusion of the shadow banking entities from the scope of the supervision has not prevented the latter from activating dealings with supervised intermediaries, and then involving the latter in the design and implementation of shadow banking operations or in risking negative consequences. It is, therefore, necessary to understand the role of banks, of insurance companies, and of funds (mutual and pension) in the shadow banking system; of course, bearing in mind that each of them is involved in different ways, resulting in exposure to different risks. Consequently, the relationship between the role of traditional operators in the shadow banking system and the shadow banking entities in the regulated market has to be addressed.

All in all, if it is clear that the orientation intended to keep the distinction between the traditional intermediaries and the shadow banking entities, then—at an operational level—the financialization of the economy and the globalization of markets puts a strain on the regulation barriers that attempt to exclude banks from the shadow credit intermediation process. Therefore, in legal terms, the relevant legal

Table 4.1 The roles in the shadow business

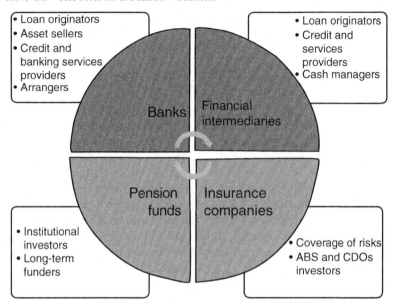

framework should provide a rule to match the aforementioned barriers (and, therefore, to include only traditional banks in the perimeter of supervision) with the reserve of public rescue only in favor of the supervised entities (the latter being instrumental to the protection of savings). Hence, there is a need to identify what are the supervisory policies that are able to achieve the result of surveying and managing its operational risks (that penalize the capital market) and, at the same time, leading the activities of the shadow banking entities to the purpose of social utility (Table 4.1 outlines the roles in the shadow business).

4.2 The role of banks in the shadow banking system

To start with, a few considerations are necessary that will be clarified precisely in the following chapters. At first glance, the shadow business of any traditional bank may seem like a contradiction in terms, an operational degeneration, an effect of the lack of internal controls or supervision. This does not cast the doubt that the same traditional definition of the phenomenon under observation leads us to rule out that these operations are realized. However, economic analyses have repeatedly identified the existence of services provided by banks to the arrangers of shadow banking operations; services that are also related to the subsets of the phenomenon named internal shadow banking subsystem and external shadow banking subsystem.[5]

Moreover, the success of the business model originate-to-distribute allows banks to negotiate in the shadow banking system the sale of a portion of their assets in view of further profit margins (or, better, higher levels of RoE) than those compatible with the parameters of capital adequacy required by the Basel Accords. It should be noted that the presence of banks in the shadow banking system can be the result of the development of *universal banking*, and the consequence of the success of the banking group or financial conglomerate business models.

It is easy to understand that, in the presence of supervised groups or conglomerates, there may be further negative externalities due to interaction between regulatory provisions that operate on distinct logical levels (i.e. the collection of savings and the joint exercise of the credit, the investment of resources of others, and the provision of guarantees versus the collection of a premium). This applies, in particular, to the risks associated with the complexity of the organization of "unitary groups" that carry out different activities (of banking, finance, or insurance type) under differentiated legal criteria (stability, fairness, and solvency).

Both the performance of services and the sale of assets do not seem to be contrary to the banking legal framework, nor able to obstruct the public supervision. Indeed, in the European Union, the credit institutions are defined as enterprises "whose business consists in receive deposits or other repayable funds by the public and to grant credits for its own account" (art. 4, para. 1, point 1, EU Regulation no. 575/2013). Therefore, this definition lacks an explicit reference to the activity of leverage or maturity transformation that qualifies the essence of traditional banking. Moreover, there are no rules providing the duty to adopt exclusively the originate-to-hold business model.[6]

This, however, does not allow us to consider that the supervision ends with the control of the deposit and credit composition, as it is extended to all activities carried out by banks (and, therefore, the exercise of any financial activity, the performance of services in connection with it, the granting of guarantees, and the conclusion of derivative contracts).

On the contrary, it should be considered that the credit institutions are subject to "holistic controls" (renewed by Directive 2013/36/EU and EU Regulation 575/2013), as a result of a process of coordination of the previously applicable national provisions, which allowed the creation of a single capital market (within the European Union), focused on the freedom of establishment and provision of banking and financial services (by the aforementioned entities).[7] Consequently, a bank that has access to the shadow banking system is still under the public control. Moreover, this access cannot open a direct "route of infection" from the shadow banking risks to the supervised industry and, consequently, cannot end up affecting depositors who have chosen to invest in the safer (because supervised) banking sector.[8]

I shall dwell on the possibility that certain contractual relationships, even if carried out on a bilateral level, allow the transition of the risks from one system to another. Obviously, there is a particularly complex network of relationships; and this makes the relevant issue difficult to assess (both at a quantitative and qualitative level).

In some cases, a bank may find it convenient to sell the riskiest exposures, in order to improve its financial situation and to transfer (to its shadow competitors) the burden of managing the operational difficulties of past-due or impaired exposures. In addition, the shadow banking entities are sometimes interested in acquiring these risky exposures, given the opportunity to practice more profitable interest rates (due to the high level of the premium for the risk of insolvency of the debtor) and, therefore, to make the most of the competitive advantages of the shadow system (with greater gains in absolute value). Consider, then, that the securitization of risky positions can align certain economic incentives (that appear able to support the establishment of long-term legal relations). This is true also for the operation where the sale transfers exposures and risks in such a way as to release in full to the bank, *cooling down* the traditional industry (and, therefore, increasing the levels of safety of savings entrusted to it).[9]

From another perspective, it is necessary to dwell on the "loan originator role" that traditional banks can play in the shadow credit intermediation process. In fact, banks are able to provide one of the best raw materials to nurture the production chain of the CDOs: the loans granted after the standard banking control procedure of the debtor. With this role comes a great responsibility for the proper assessment of creditworthiness, which a bank cannot make in a short-term perspective (related to the limited period of holding of credits), as it must be aligned at the (long) timing of future securitizations and re-securitizations.[10]

Of significant importance is the regulation that requires banks to proceed always to the verification of the scoring and credit rating of the customer, even if they operate under the originate-to-distribute business model. Hence, we feel the need to verify whether the intermediaries who originated the loans are organized in such a way as to better manage the information gained when they drive the loan agreements (on the likelihood that the borrower will default on its debt obligation).

Hence, the role of banks participating in the shadow credit intermediation process as loan originators is very important, because their activity corresponds to the collection of information that—sometimes by way of summary—will be available to those who organize the shadow process. In other words, *originating banks* are responsible for the selection of the assets that give content to the entire process and, therefore, the quality of the elements underlying the asset-baked securities. Furthermore, they set the prices for the "sale of loans," formed on the basis of the (qualitative) aspects considered at the time of assessment (of creditworthiness). The same is true also for the case in which the money is granted on the basis of external ratings, since in this circumstance the cost incurred by the bank (for the purchase of the evaluation) may not be reimbursed by the transferee.

The position of the banks that carry out operations of self-securitization (or retained-securitization) is different. In this case the operation ends without the financial instruments being sold to the market (but retained by the banks themselves). There is no doubt that such an operation does not fall outside the rules of supervision, given the irrelevance of the same towards the definition of the risk profile of the banks (which, logically, should be weighted by transparency, given the risks of the assets transferred and not those of the securities issued). Otherwise, the effect would be to admit an operation that, for the result that it produces, is contrary to the principles of safe and sound management of credit institutions and the global regulatory framework for more resilient banks and banking systems.

Other roles may be carried out by banks within the shadow, according to the possibility of providing "support measures" (or rather, guarantees) and facilities in the placement of the final securities. These roles do not suggest conclusions that are very different from what the ordinary supervisors would expect in the monitoring of the traditional banking activity. But, although the fundamentals have remained unquestioned, the difficulties (suffered by banks) in understanding the counterparty risks (given the aforementioned nature of the shadow entities) must be taken into account.[11]

Finally, the critical issues related to competition between these two systems must be considered. There are no grounds for believing that banks are under pressure because the shadow banking entities are not subject to the burden of capital requirements stated in the rules of prudential supervision. Consequently, there is a prospective reduction in prices, which will force banks to recuperate—in some way—the expenses that they incur as a result of the aforementioned asset weights (in order to avoid losing a substantial share of the market in favor of the shadow banking entities).

Obviously, there are effects related to the achievement of an equilibrium price (in the capital market), since these reduce the gross income of the banks. Hence, the decrease of the relative levels of profitability, useful results, and, therefore, the self-financing capacity of the same (which is related to the ability to accumulate reserves in a position to increase its allocation of regulatory capital).

Concluding on this point, it must not be forgotten that the presence of banks is qualified by the activities that take place within the shadow process. It seems, therefore, possible that the actual configuration of relationships (contractual or participatory) raises issues that are not always adequately assessed by the current system of regulation. Hence, the interest for the introduction of new rules able to weigh the effects of the shadow operations with the volume of the activities and, therefore, to continuously monitor the capacity of the managing organization to ensure the safe and sound management of the firm.

4.3 The dysfunctions of internal controls and weaknesses of other safeguards

It should be obvious that the banks' access to the *shadows* is compatible with the banking system only if it is contained within the risk constraints provided for by the applicable prudential regulation. Consequently, the system of internal controls shall be responsible for ensuring that intermediaries realize (or, rather, become partakers of) hazardous operations only in order to improve the quality of their balance sheets or increase the scale of their profits.

In this context, it will be up to the system of internal controls to prevent the presence of banks in the shadow banking system becoming a solution aimed at externalizing risks (through the securitization of assets with poor quality) or to pave the way for a market in which it becomes possible to carry out particularly speculative operations.

Therefore, the need to consider jointly the operation of the shadow banks and the need for a strengthening of internal control systems is understandable. This applies, in particular, to the recent interventions that, in line with the guidelines published by the European Central Bank, redefine—qualitatively and quantitatively—the overall structure of the model of banking governance.[12] These actions proceed to intensify banking and financial supervision; in line with the supervisory policies forming the basis of the EBU (and, therefore, the new regime of control over intermediaries of systemic importance).[13] However, the problem of how to reduce the abovementioned operations is still unsolved without the adoption of safeguards designed to prevent banks leaving market-based operations to prevail on the traditional one.[14]

That said, the question should be the role of the compliance function in the governance of banks that operate in the market-based financial systems. This function, in fact, should not exhaust its tasks in the fight against the adoption of illegal behaviors, but must also prepare a set of internal rules (and control self-assessment processes) designed to give practical effect to the safe and sound management (and, therefore, to ensure the quality of intermediary activities) in any business of the credit institution.[15]

It is my belief that access to the shadow banking system is prevented by the malfunction of internal audit function, but that it is driven by precise management decisions taken by the executive leadership. Hence, the need for assessing whether the setting of the internal control system (as a whole) identifies an appropriate instrument to ensure the proper execution of the abovementioned operational choices. It goes without saying that such a system may result in limitations of the business and, consequently, becomes a prerequisite for the introduction of barriers (in addition to those provided by primary legislation) between the regular course of banking activity and the shadow operations.[16]

In this context, the "principle of prudence" shows its central role, according to which the safeguards posed by internal controls should prevent banks from participation in transactions that—in subjective or objective terms—exceed the limits set by public oversight. It is hardly necessary to point out that the current regulation ascribe to any intermediary the responsibility of defining its risk profile (so it is up to it to ensure the right balance between equity and loans). Therefore, the concrete organizational solutions adopted by the management will have to avoid degenerations that, in a concrete form, project traditional banking outside the regulated market.

In addition, the proper exercise of the compliance corresponds to a limitation of the activities of the banks in the shadows, which could be constrained within the role of loan originator and service provider. Only adequate collaterals can allow the bank to support a stage of the shadow credit intermediation process.

In this way, a new role for the banks arises, which is more balanced in terms of profitability and prudence. It appears consistent with the cycle of innovation that characterizes the banking industry and the circuits of finance. It is clear, in fact, how the systems of administration and control (of the banks) should ensure the possibility of providing a range of services adapted to the needs posed by the evolution of operational techniques. Without it, this leads to the assumption of excessive risks or to the arrangement of inappropriate relationships with individuals operating outside the scope of supervision.

In other words, banks' internal controls should not only assess the legality of the provision of services to the shadow banking entities, but they should also prevent the search for new and more profitable brokerage circuits leading to operations that go beyond the prudential rules.[17]

Ultimately, if this analysis raised certain concerns with regard to the presence of regulated intermediaries in the circuits of the shadow banking system, it must now be pointed out that the trends of the industry confirm the compatibility of this phenomenon with a prudential approach. In the end, the aforementioned provisions (on internal controls) are clearly intended to introduce new, additional safeguards designed to prevent the mingling of bank-based intermediation with the market-based one. The aim of this is also to limit the effects of the risks that arise in the shadow banking system, preventing the transfer within the economies of the banks (and in this way towards the public's savings).[18]

4.4 The action of insurance companies

The activity of insurance companies in the shadow banking system may appear inconsistent with the forms of supervision to which such companies are subjected. Considering this carefully, even their participation to the shadow credit intermediation process (by providing services and hedging) does not seem to fully comply with the definitions given in Chapter 1.

It is easy to remove the doubt that we are in the presence of supervised institutions, which realize their business within a regulated environment (such as the one in which the resources reserved for the coverage of risks of various kinds, including longevity, social security, and welfare circulate). I am talking, therefore, of subjects who are controlled on the basis of their involvement in the processes of asset management according to criteria suitable for assuring their solvency over time (because of the reversal of the production cycle that characterizes their typical activity). After all, this is the background of Directive 2009/138/EC (on the taking up and pursuit of the business of insurance and reinsurance, so-called Solvency II Directive) and Directive 2013/58/EU (postponing the application date of Solvency II).

It is, therefore, necessary to investigate the participation of insurance companies in the shadow credit intermediation process, as evidenced by international economic analyses, in order to understand how the dual role that these companies perform within the shadow banking system should be taken into account (by the supervision authorities). Indeed, they can participate both in the aforementioned process (providing insurance services or derivative contracts) and subscribing CDOs offered on the market (in order to invest their cash). Obviously, it should be kept in mind that, as it has been clearly stated by the FED, these subjects do not have access to central bank liquidity and, therefore, can be more easily assimilated—from a monetary point of view—to shadow banking entities.[19]

We shall also assess whether the organizational structure provided by the applicable legislation is able to ensure the proper performance of guarantees in favor of the shadow banking entities and, therefore, if it allows correct pricing of the required "premium," as well as investing it in "technical reserves" that will retain their value in the event of the occurrence of the insured risk.[20]

This approach may appear to be aimed at verifying a conventional type of insurance's issues; this is because, frequently, these firms provide their services to parties who operate in markets that are not subject to financial supervision (as in the case of insurance provided to industrial companies). On the contrary, insurance companies' access to the shadow banking system poses new types of problems. In particular, because in the shadows there is the assumption of risks related to financial effects of certain operations, such as the possible negative outcomes of the expansive performance of the economic-monetary variables (and these effects are not due to extraneous events in the capital market).

It goes without saying that the insurance companies enter into a relationship with shadow banking entities, which by their nature present a particular company risk (and, therefore, of counterparty), due to:

- the environment in which they are incorporated;
- the characteristics of the organizational and operational structures; and
- the internal variability.

Therefore, in the shadow banking system, these entities will need to provide *innovative* strategic lines of action to deal with new risks (of dominance, flexibility, and integration). This suggests the need for an adequate corporate structure, able to support a business formula different from that of the traditional matrix.

By way of summary, it is clear that the organization of insurance companies must avoid the procustion of a shift of risks from the shadow banking system to the insurance market. There is no doubt that the presence of these companies cannot be resolved in a role of mere guarantee, such as to ensure that an alternative operation is provided in the event of default of a shadow banking operation. Otherwise, this could mean a market mechanism that requires the direct transfer of resources from the shadow banking system to the insurance market, through the payment of fair premiums (because of the prudential setting that qualified the computation rules of the latter).

In other words, the participation of insurance companies (in the shadow process) may produce interactions that transfer from one system to another both the incidence of risk (systemic or not) and a portion of the profits that market-based financing should achieve (in competition with the banking sector). In this context, the need for a form of supervision intended to display the morphology is identified, along with the strategic guidelines of action that must characterize the governance of regulated companies (which provide their services to the operators of the shadow banking system). That is, in order to pursue—even in the insurance sector—the objectives of (market) stability, of prudent management (of the operators), as well as a (general) sustainable economic growth.

On this point, the recent structure of European regulation indicates new forms of intervention, which are based on a more careful monitoring of the access and of the pursuit of the activities of insurance and reinsurance. A new legal framework corresponds to the adoption of Directive 2009/138/EC, which allows companies to consider a simplification in the coverage of the risks and commitments (*Recital* no. 2); and this is in view of a proper functioning of the market (*Recital* no. 3). However, these new rules do not succeed in effectively managing "captive insurance and reinsurance companies" and, therefore, it is not given to find in this case a discipline that safeguards the companies that only cover the risks associated with the group of shadow arrangers to which they belong.

In addition, the presence of new solvency requirements is linked to the efficient allocation of capital moving within the EU internal market.

More generally, the options available to the European regulator do not solve the issues related to the nature, scale, and complexity of the obligations that are required to ensure the activities taking place during the shadow credit intermediation process. Indeed, the European authorities felt the need for a moment of reflection prior to the application of the amendments to the European insurance regulatory framework, given the aforementioned adoption of the Directive 2013/58/EU amending Directive 2009/138/EC concerning its transposition deadline and the date of its application.

With particular reference to the subscription of CDOs by insurance companies, it is important to highlight the contents of the delegated regulation of October 10, 2014, bearing new rules for the evaluation of assets and liabilities, own funds, and solvency capital requirements.[21] I refer, in particular, to the changes that the insurance companies will have to make to their organization in order to participate in the securitization transactions (so-called "special requirements for high quality securitization").[22] Therefore, currently there are requirements that relate the amount of capital requirements to the actual assumption or translation of risks. Hence, insurance companies need to organize themselves keeping in mind that they "can only invest in securitization positions where originators, sponsors or original lenders retain a material net economic interest in the underlying exposures."[23]

Even when only involved as investors (and, therefore, without participating in the shadow banking operations), these companies must address specific problems arising from the entrance of a supervised company in an environment free from controls (as the shadow banking system is). The selection of the securities in which the individuals in question may invest their cash, in fact, come into consideration. Hence, the need to verify "continuously" (i) the suitability of the CDOs to be included in the assets of the company (also in order to proceed to the coverage of technical reserves), and (ii) the compatibility with the needs of a prudent management (to be assessed, in the sector under consideration, with regard to the stability, efficiency, competitiveness, and proper functioning of the insurance system).

It is easily shown that the business function most affected by the new rules is the one that deals with the management of the treasury, because of the peculiar "maturity structure" associated with the financial instruments in question. It will be the task of the cash pooling managers, therefore, to provide reliable actuarial valuations, according to the peculiar responsibility of the boards of directors (called to select the titles that have risk profiles compatible with the obligations taken and with the safety, profitability, and liquidity standards set by the regulator). These are, of course, responsibilities related to compliance with the regulation designed to *stem* the assumption of risk positions (that would be difficult to bear by the company itself).

Obviously, if these investors intervene in an intermediate stage of the shadow credit intermediation process, the abovementioned business functions will need to pay more attention to the investment choices in "long-term MTNs and bonds"(or shares issued by the shadow banking entities), to which are obviously associated higher levels of the sector's risk sharing.

That said, it should be emphasized that the intervention of the insurance companies can induce operators to perceive that "credit quality is driven by the 'credit-risk free' nature of collateral that backs shadow bank liabilities, as it is often enhanced by private credit risk repositories."[24] Conversely, this is not always true. It must be considered that the ambivalence of the position of these companies—in respect to the shadow banking system—introduces a new kind of danger, which is related to the possibility that parts of the technical reserves (placed to honor the guarantees given) are invested in assets exposed to risks associated with the performance of the shadow banking system (either because they constitute titles that refer to it, or because of the presence of systemic interconnections). Hence, there is the possibility that—in the case of a negative trend of the phenomenon under observation—the abovementioned activities are subject to a "devaluation spiral" and, after a while, they become insufficient to support the delivery of the promised performance.

It is evident, therefore, how the presence of insurance companies in the shadow banking system cannot be considered free from negative implications. In fact, it poses specific risks against which new organizational structures must be provided. This is particularly true with regard to the role of the "key functions" referred to by the mentioned Directive 2009/138/EC: actuarial, risk management, compliance, and internal audit. It will then be the task of the European Commission to indicate—on the basis of the *Green Paper on the Shadow Banking*—the consequences of shadow operations, which certainly may result in "deviations in their risk profile" and, therefore, in "the elements to consider in deciding on an extension of the recovery period for undertakings having breached their Solvency Capital Requirement."[25] This happens in a context in which primary emphasis should be given to reporting and public disclosure procedures, which shall be adequate to the high information required for safe performance of the financial operators within the shadow banking system.[26]

Concluding remarks will follow the identification of a new field of action for the public authorities, in which the regulation of the governance of the insurance companies should be aimed at avoiding pro-cyclical trends and, more generally, the presence of links of systemic relevance. Hence, the hope that EIOPA will be able to take into account the features of the shadow banking system when it will be required to identify higher (organizational and operational) standards for "improving the functioning of the internal market, in particular by ensuring a high, effective and consistent level of regulation and supervision taking account of the varying interests of all Member States and the different nature of financial institutions."[27]

4.5 The involvement of pension funds

Worthy of attention are the subjects that have access to the shadow banking system in order to invest their assets in the securities issued by the shadow credit intermediation process (ABCP, ABS, and CDO). These subjects provide the flow of money that finances the whole circuits, and then they suffer the risk of the system's insolvency.

Individuals and institutional investors can be placed in the same position, but only the latter are required to perform a market activity giving effect to specific investment policies (which are left to the free determination of the asset managers, but have been predetermined by third parties and approved by the supervisory authorities).

Therefore, certain institutional investors (and other asset pools meant to generate stable growth over the long term) must be taken into account, since the presence of pension funds has a quantitative impact on the overall financial position of the shadow banking system. It goes without saying that the high return of shadow securities raised the interest of these funds, them both being CDOs (that meet the needs of the relevant cash pools), or other intermediate assets with longer duration (ABCP and ABS). It is surely not a coincidence that this kind of subject has been described as "long-term funder," that is, "its assets are the ABS and the (unsecured) corporate bond, and its liabilities are the pension obligations to the nonfinancial sector." A particular bargaining strength follows, resulting from the fact that "a pension fund may engage in frequent trades and may choose to quickly dump assets that it no longer wants."[28]

Following on from this approach, it may also be agreed that the presence of pension funds represents an element of stabilization of the system, since their financial mechanisms "typically do not rely upon short-term funding and are generally not considered runnable."[29]

It can be assumed that a similar effect could influence the conduct of the operators, given that the investment policies of pension funds, on the one hand, can influence the choices of those who organize the shadow banking operations (in order to create financial instruments adapted to meet the demand) and, on the other hand, expose the invested assets to the shadow banking risks. It appears, therefore, necessary to place these subjects at the border between the traditional and the shadow systems.

To summarize: although pension funds belong to the traditional financial system (and, therefore, are subject to public supervision), the investment of their portfolios in the shadow banking system (and its exposure to the risks) raises issues similar to those applying to the abovementioned shadow banking entities.

The techniques described here affect the arrangers of the shadow banking system, which shall take care to prevent any risk on their behalf by engaging preliminary relations in which they involve future investors. Hence, the aforementioned funds shall implement their investment policies in order to simplify the management of the structured financial instruments; mainly because of the link with the real economy that is often searched by more cautious professional asset managers.[30] This is why high levels of flexibility for the management may allow the construction of shadow banking operations *tailored* to specific investment policies; operations that are expressed through the issuance of financial instruments and able to collect both the resources owned by pension funds to provide social security benefits and the cash sums held by them for treasury requirements.

In this context, we should keep in mind the rules that govern the possibility of entrusting the financial management of pension funds to third parties (in the forms provided by the rules of harmonization introduced by Directive 2003/41/EC). This is one of the legislative safeguards aimed at avoiding participation of such funds in the shadow banking operations affecting the *par condicio* between financial operators, as provided by special regulation. In particular, with regard to the operational correctness, the affixing of capital requirements, as well as the prediction of general investment restrictions and conflicts of interest.

There are, therefore, subjects who, although supervised, cannot be considered unaffected by the aforementioned shadow effects. They are influenced by investment in ABCP, ABS, and/or CDO. Therefore, their external character does not allow the consideration of pension funds as *strangers* to the shadow credit intermediation process. This is true also when considering that pension funds' demand characters interact with this process, being able to influence the structure of ABS (and, therefore, its (market) price). This has obvious consequences for the accounting information produced by the application of fair value (as it is accounted a higher value than that attributable to the underlying assets).

In the end, the proper definition of the internal organization of pension funds, therefore, corresponds to the ability to select the best stocks and, hence, to support the production cycles (of CDOs) more reliably (promoting a process of selecting operators based on the criteria of operational quality). Thus, the importance of an effective supervision on the "investment process" and the proper configuration of the internal functions of compliance, risk management, and audit is concluded. There is, indeed, force in the argument that these conditions are similar to those that motivated the European option to establish a uniform system of prudential supervision of institutions for occupational retirement, organized on a continental scale (especially after the establishment of EIOPA). Consequently, inferring these conclusions from the discussion in this chapter would not misinterpret a supervision that—in controlling the investment policies of pension funds and the quality of their assets—produces a positive externality for the shadow banking system, which will have an incentive to offer transparent and safe securities.

5
Shadow Banking Operations

This chapter chiefly concerns the analysis of the shadow credit intermediation process, but the main purpose remains dealing with difficulties of the credit transformations, in order to understand the allocation of risks and benefits among the shadow banking entities.

Thus, I cannot achieve this goal without taking into account the EU and US rules governing the techniques of securitization, and their impact on the market-based financing. It is necessary also to analyze the role of "credit ratings," before considering the problems related to the offering of ABCPs, ABSs, and CDOs. Furthermore, I must analyze both the (general) use of derivatives and the (specific) management of national states' sovereign debts in the shadows.

The importance of these matters cannot be underestimated, even if the global networks of financial regulators seem intent on *monitoring* the shadow banking operations in order to increase the transparency of the financial market.

5.1 Shadow credit intermediation process

The operations activated by the shadow banking entities give content to a process aimed at creating securities that, as we have seen, support the circulation of capital from those with a surplus to those in deficit. So far, these operations have been considered as a whole, being indicated by the abbreviation of "shadow credit intermediation process." However, these operations involve a *multitude of affairs* connected together by specific purposes (i.e. credit enhancement and maturity transformation). Thus, it is important to be aware that the shadow banking system does not qualify for the pursuit of a unified economic business, but for the interconnection of a systematic series of activities (differentiated by object and subject).[1] Moreover, as will be seen, something suggests that this process should consist of a series of contracts made by a single financial intermediary (i.e. the hypothetic figure of shadow bank).

In legal terms, the shadow credit intermediation process does not occur in an integrated manner under a sole entity. This is one of the most important features that make the shadow banking system different from traditional banking (in which all stages are realized within a single intermediary). To this feature correlates the interpretation according to which "some shadow banks are businesses, not funds or vehicles."[2]

The dynamic and sequential nature of the shadow credit intermediation process (highlighted by the aforesaid analysis of several international authorities) will be taken into account. It corresponds to a legal framework in which the loans to the real economy (in the beginning) and the financial securities (at the end) shall be considered, on the one hand, as assets that will be subject to a credit-transforming process (carried out at each stage of the process) and, on the other hand, as the "initial-final product" to be produced in order to remunerate all those who have executed these (shadow) operations (see Table 5.1).

Table 5.1 Shadow credit intermediation process and regulated activities

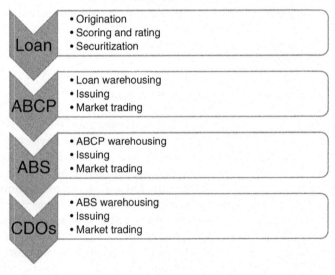

As a result, the need to control this process, from one stage to the other, allows the implementation of a profitable "wholesale funding" channel; this can also balance the development of the entire shadow process (and, therefore, the defense of the legal positions to which correlates a need for protection).

In any open market system, the mechanisms of credit intermediation (and then also those of the transactions carried out by the shadow banking entities) can influence the financial stability. Thus, it can be hypothesized that the supervision must take into account the whole shadow process, rather than just check (and face) the negative consequences on the capital markets of certain operations (involving banks and shadow entities). In fact, the public control of these effects can avoid damages to credit institutions and financial firms, which work (too) in close proximity to the shadow banking system.

Hence, there is a need to establish what are the interventions that can manage (negative) externalities produced by shadow banking operations in the global financial system (through interconnections with the economies of other subjects acting in the capital market).[3] And, again, it appears to be necessary to proceed with an examination that, keeping in mind the holistic importance of the shadow credit intermediation process, dwells on the analysis of the legal framework of the contents of the operations, because—as shall be seen—these externalities can produce micro- and macro-prudential risks.

Therefore, the interest is focused on the rules of the relevant market (for shadows securities) and, with them, on the practices regarding the composition of mutual interest takes place, as well as the transfer of resources. These rules and practices are the basis to understanding the way an operation is structured and a contract is negotiated to meet the demand—for liquid financial instruments—that feeds the shadow banking system.

Using this logic, I shall refer to the results reached by the European Commission, which underlay a definition anchored to two intertwined pillars (i.e. entities and activities), with a specific emphasis on the operations carried out by "special purpose entities" (which exercise an activity that can create liquidity and/or transform the deadlines, including within them both securitization vehicles such as ABCP conduits and SIVs, and other types of SPVs in use in international practice).[4] These results are linked with variegated, but related, problems: from the worthiness of the economic activity exercised, to the weight of the limits to private freedom, to the need for the attribution of responsibilities.

Therefore, the wide freedom recognized in financial markets for self-regulation of the "business relations" that give content to this shadow process comes into consideration. The chapter shall then move on to analyze the content of the relevant transactions, taking into account that these results form the mere consent of the parties, without being costumed to predetermined patterns or models given by the supervising authorities. Hence, there is the need to focus on the innovative nature of the shadow operations, as the private autonomy has few limits in the system under consideration.

5.2 Undetermined (contents) and unconfined (boundaries) of shadow banking operations

Beginning the analysis of the shadow credit intermediation process, this section shall move from the analysis of the operations of credit enhancement and maturity transformation, and then verify the concrete possibility of sustaining the supply of (long-term) credit over time, through the offering of (short-term) financial instruments. The same has to be said with regard to the techniques used in order to take advantage of the economic benefits of financial leverage (even through the activation of credit lines by traditional banks or insurance companies, as provided by the recent Italian legislative decree no. 91/2014, converted into law no. 116/2014). Hence, specific issues will be addressed regarding the shadow banking operations that are completed through the negotiation of derivative financial instruments (aimed at improving the quality of the securities offered in the market).

Any shadow banking operation—in pursuing the goal of connecting the supply of loans (to the real economy) and the demand of financial instruments (of cash pooling managers)—generates positive effects for the capitalist system, but also risks and costs. Any of these operations brings also benefits, both at the macroeconomic level (by introducing a form of alternative banking and then improving the competition in the market for loans) and at the microeconomic level (as they allow the taking and managing of risks other than those permitted under the prudential rules).

It goes without saying that, following the great crisis of this millennium, attention is focused on the possibility for generating systemic risks that extend to affecting the performance of the whole financial market or the results of the monetary policies.[5]

In this context, "securitization" and "wholesale funding" represent a sort of *paradigm* useful for structuring operations that, in the end, allow the incorporation of a series of sub-operations. Consequently, these operations require, on the one hand, banking, financial, or insurance services and, on the other hand, derivative contracts. It can be understood that any shadow banking operation requires a "daisy-chain of non-bank financial intermediaries in a multi step process," but it does not exclude the participation of supervised companies.[6]

In particular, the description of the shadow banking system proposed by the FED should be kept in mind; which—as we have seen—specified the "economic substance" of these operations, by highlighting the peculiar configuration of the CDOs that are offered in the market in order to satisfy the demand for safe and short-term securities (other than the solutions offered by the banks).[7]

Obviously, the underlying loans (to the real economy) follow the traditional rules provided to set up the "fundamental legal relationship" between the lender and the debtor. This relationship regulates the repayment of both principal and interests. Although the shadow credit intermediation process calls for the subsequent sale of the credit arising from such loans (from the lender to a special purpose vehicle or an—alternative or not—investment fund), this relationship guarantees also the reliability (and creditworthiness) of the assets that will back up the following securities.[8]

To this initial transfer corresponds, then, a number of securitization transactions involving a variegated set of loans (provided by different originators). It is apparent that, in principle, the procedure for the acquisition of assets by an SPV takes place in the relevant market and, therefore, it shall follow the rules that allow the operators to choose the credits, bearing in mind that the latter shall be used to support the issuance of an ABCP, before, and ABS or CDO, after.

It goes without saying that the successive stages of re-securitization of financial instruments allow, then, the adding up (in a second SPV) of ABCPs issued through the use of different credits. Hence, the possibility of a "pyramidal" development of these operations shall also be considered. Then, as shall be seen, at each step, the increasing complexity of the operation makes more and more difficult the understanding of the whole intermediation process and, therefore, the assessment of the risks underlying these securities (and consequently the allocation of operational responsibilities).

In light of the foregoing, it is evident that the shadow credit intermediation process is composed by a sequence of operations that are not independent, but linked by a logical or legal pattern, due to the practical choices made in any previous stages. Thus, the arrangers have the freedom to choose the contents and the boundaries of any securitization project (and, in general, the price or earning of the CDOs offered to the investors).

According to these practices, the risk assessments (and the risk management) require evaluation during each step of the shadow credit intermediation process, and then any contract or SPV involved in the relevant shadow banking operations. This is the safer way to verify the correctness of the structure of these transactions, the transparency of the economic conditions, the symmetry of the benefits, and the preservation of the money invested in the shadow banking system. Perhaps, this is the only way to overcome the *impediments* related to the opacity of such negotiations and, therefore, to take advantage of the possibilities that the shadows are able to confer to the market.

Besides, the lack of regulation (or, rather, standardization) of the shadow banking operations avoids the dissemination of market practices, with obvious difficulties in the identification of safeguards that can manage the (positive or negative) externalities of such a process (including those relating to monetary issues).

It is necessary to dwell on the contracts that give legal form to any stage of the shadow credit intermediation process that follows the initial transfer of the (originated and distributed) credit. First of all, consider the type of relation that links the shadow banking entities and the companies belonging to the supervised (banking, financial, or insurance) industry. Indeed, the lawfulness of such a relation (and of its obligations) exposes any supervised entity to the negative consequences of the breach of the obligations (or, rather, to the relevant risks). To this corresponds the necessity to use precautionary provisions (and then to introduce specific and additional capital buffers for banks and insurance companies) in order to compensate the shadow banking risks (and then to avoid a financial contagion or a domino effect from the shadow banking entities to the supervised companies).

Therefore, it shall be considered here that the shadow credit intermediation process seems to be based upon contracts negotiated for:

- granting credit facilities to the SPV (by independent intermediaries that are not linked to the arrangers of any operation and, therefore, potentially excluded from any duty to consolidate the risk of the SPV[9]);
- providing counseling services in the organization of these operations (by advisors, merchant banks, and, in particular, investment bank divisions of supervised companies); and

- guaranteeing the repayment of the capital invested or an alternative performance in case of default of the process (by the intervention of an insurance company).

Even in the absence of further elements, these contracts create a "direct link" between the the shadow banking system and the supervised system (through the above credit facilities, services, and guarantees). This link represents the exposure of the whole capital market to the risks of shadow banking operations, especially with regard to the negative effects of any "maturity mismatch" or other errors in the design of the shadow credit intermediation process. Consequently, it is necessary to highlight the macro-prudential importance of any shadow banking operation that involves a supervised entity, and the supervision authorities shall verify whether "the draw-down rate assumed in the Basel 3 Liquidity Coverage Ratio should be higher for committed lines to financial companies than for lines to non-financial companies."[10]

It can now be concluded that the shadow credit intermediation process can build an unsafe connection to the supervised market, through operations placed without the safeguards usually applied to protect the "weaker parties." The shadow banking system provides a framework where an "equal negotiation" is possible and does not require any specific constraint (or covenant). Therefore, only the professionalism of the arrangers can ensure the fairness of the negotiation and the adequate exchange of information (in order to ensure the proper knowledge of these operations). It cannot be accepted that this way of contagion is not subject to specific controls to verify (i) the transparency of economic conditions, (ii) the symmetry in the distribution of benefits, and ultimately (iii) the cost-effectiveness of the shadow banking operations in which a bank (or rather depositors' money) is involved.

5.3 The operations of the credit transformation

It is necessary now to dwell on the shadow banking operations that carry out the transformation of maturities in ways *alternative* to the ones used within the traditional banking system. Implementing these operations, the arrangers try to improve the return on their equity through the enhancement of the quality of the operation activated in the shadow credit intermediation process. This explains the need to break down any offering in different tranches (of distinctive standing) or the presence of third-party guarantees, or the subscription of derivative contracts (to cover the default risk of the underlying assets).

In particular, I shall consider that the operations that realize "maturity transformation" (and, therefore, allow the use of short-term deposits to finance long-term loans) have a specific ability to increase the levels of market liquidity (when assuming rollover and duration risks). A similar assessment should be made with regard to use of "liquid instruments to fund illiquid assets," which corresponds to the effect of "liquidity transformation."[11]

In this case, it cannot be admitted that the use of banking services (and the arising obligations assumed by banks) are not regulated and controlled.[12] Conversely, the idea is shared that market transactions (even if they are not standardized) can be executed in any trading facility, given the required level of transparency (required for the smooth operation of the financial market).

It is apparent, indeed, that the regulation set for any initial public offering (of ABCPs, ABSs, and CDOs) is an essential prerequisite to guarantee the correctness and the accuracy of the improvements resulting from the sequential structure of the shadow credit intermediation process. Only the transparency of the shadow banking operations allows the knowledge of the positive effects of credit transformation—even if reached "through the use of third-party liquidity and credit guarantees, generally in the form of liquidity or credit put options."[13]

However, no adequate form of supervision on the operations of credit transformation are seen to be in use in the shadow credit intermediation process. There is a need for the aforementioned purpose of controlling the effects of the shadow banking operations to soon be consolidated within a well-defined regulatory framework, consistent with international guidelines (and then able to ensure the effective market stability, the reduction of risks, and, ultimately, the sustainable development of the real economy).

5.4 New securitization techniques

In this and the following paragraph the focus will be on certain perplexities arising from the shadow credit intermediation process. Consider the (neglecting of the) idea that the shadows do not hide a network of securitizations and re-securitizations aimed at avoiding obligations arising from the regulatory framework. Thus, the following considerations will help us in pursuing our main theme and then in analyzing a complex system that is (itself) an *externality* of the legal effects deriving from the freedom of providing services and the implementation of credit channels alternative to those of banking.

Consequently, the discussion moves from the funding strategies, that is, the offering procedures of financial instruments (or of participation shares in the case of investment funds).[14] On this point, it is important to take into account the will of the European regulator to monitor the activities of the "financial vehicle corporations" engaged in securitization transactions (Regulation (EC) no. 24/2009 of the ECB), in order to gather information concerning the close links between these securitizations and the operations of the monetary institutions. This is a supervision of the information, data, and news related to the trends of the market, but this intervention cannot influence or regulate the latter. It seems that the ECB aims to learn the fundamentals of this sector, and then—as will be seen—there is still only a limited solution (to the problems arising from a process not always executed in a transparent way).

It must be kept in mind that the US legal system—and, in particular, Section 941 (c) of the Dodd-Frank Wall Street Reform and Consumer Protection Act—had already considered the problem of upholding a direct interest (of the arrangers) in this kind of operations. This solution, however, was adopted to align the market practices to the effects of the "new risk retention requirements" to be developed and implemented by the federal agencies, and of Statements of Financial Accounting Standards nos. 166 and 167 (FAS 166 and 167).[15] The approach adopted by the Italian legislator can also be taken into account—in the law decree no. 145 of 2013 (as amended by the Conversion Law no. 9 of 2014)—in regulating the same topic, providing that banks intervening in any operation of securitization must hold a direct interest in the same, due to the acquisition of a certain amount of the financial securities (ABCPs, in the face of which, of course, a *bit* of the credit institution's regulatory capital must be committed).

Undoubtedly, the aforementioned regulation tries to improve the safety of the securitization by involving the originating bank in the insolvency risk. Consequently, it is clear that these rules realize a mixture of typical banking activity and shadow credit intermediation process. However, it seems that the mentioned (American and Italian) regulators come to this solution on the basis of a different logical orientation. Indeed, in the first case there is the intention to decide on the "discussion of the economics of securitization, a summary of the underlying collateral, and differences in the securitization 'chain' linking originators to investors";[16] whereas, in the second, the aim to achieve an increase in the quantitative data (credit supply) to support the demand of enterprises and boost the economic recovery.[17]

Moving on to consider the position of the Bank of England on this topic, it shall be highlighted that the last (together with the ECB) attributes special importance to a *specific* interpretation of the "state of play in the securitization market," which draws the initiatives relating to the regulation of the establishment of the so-called "G20 retention principles" in order to complete the set of safeguards (which are summed up in the capital and liquidity requirements of the Basel Committee, as well as in Solvency II).[18]

There is, therefore, a common trend towards an "interventionist policy" that can result in an "aim of ensuring a more resilient shadow banking system internationally, including securitization markets."[19] It remains to be verified, however, whether the results of this approach will be compatible with the European regulatory framework and its new system of financial supervision.

Relying upon the Italian experience, it can be hypothesized that "banks can effectively counter the negative effects of asymmetric information in the securitization market by selling less opaque loans, using signaling devices (i.e. retaining a share of the equity tranche of the ABSs issued by the SPV) and building up a reputation for not undermining their own lending standards."[20] It goes without saying that such a hypothesis can suggest the immediate conclusion that "securitized loans have a lower probability of default, indicating that the securitization market can provide an appropriate tool for transferring credit risk efficiently."[21]

However, it is apparent that the abovementioned regulatory structure tries to avoid a misalignment of the incentives underlying the selection of the credit, which arises from the fact that the costs (of this evaluation) are faced by the supplier-seller (and not by the transferee or by other parties that will suffer the damages arising from the failure of the debtor). Then, it can be understood that the regulator faces the actual risk of a search for increase in volume that can reduce the accuracy of the relevant selection. Nevertheless, this is an issue addressed by Italian scholars in recent times, who have argued that "the crisis is generated by the securitization of credit risk," or, if it is possible to reformulate this sentence, that the crisis was caused by *the absence of a prohibition of* the securitization of credit risk.[22]

All in all, the current banking regulatory framework denies the possibility that any shadow banking operation will be able to fully transfer the risks arising from the securitized loans to the shadow banking entities and by them to the final investors, because of the retention requirements provided for the originator.[23] Therefore, one expects an action plan aimed to restate supervision practices on securitization markets, taking into account the new structure developed by the financial engineering and the related riskier transactions.[24]

5.5 Liquidity, maturity transformation, and financial leverage

Everything gets more complicated when one considers that the most recent shadow banking operations combine the financialization of credits with (i) the transformation of maturities, (ii) the use of leverage, and (iii) the assurance of minimum levels of liquidity. And this happens in an opaque environment, in which the description of the overall level of quality is entrusted to the granting of a summary assessment by a credit rating agency (as will be seen in the following paragraph).

This suggests that the application of the aforementioned techniques to the shadow credit intermediation process seems to be in contrast with the originate-to-distribute credit intermediation model, where it should be possible to transfer to third parties the negative effects of credit default (even if they are due to the inefficient creditworthiness evaluation, in terms of a bank's capacity to select reliable borrowers).

It goes without saying that the experience of subprime mortgages has, in some ways, marked a point of view that may affect the judgment of any interpreter.

There is no doubt that the construction of serial chains—in which the risk transfer is progressive and uncontrolled—allows individual episodes of insolvency (at some point in the chain) to cause one or more domino effects, which may qualify the same way as side effects not directly related to the insolvency trigger. At the same time, consider that the shadow banking operations did not only allow the conversion of loans (to the real economy in instruments required by the financial market), but also had a favorable impact on the *liquidness* of the originators, providing a solution to the banks that—in the exercise of their own business—were not able to manage the related liquidity risk.[25] This was possible until the prudential regulation in force failed to provide particular requirements on this subject.[26] It mustn't be forgotten that (at micro level) securitization remains, at least in a first stage, an effective risk management technique.[27] A similar consideration cannot be made, however, with regard to the macro effects, because economic studies do not show unequivocal results, compared to critical positions of public supervisors.[28]

That said, the regulation will have to take into account the securitization made within the shadow banking system also referring to their impact on liquidity levels of the capital market.[29] Of course, their impact depends on the legal framework of the operations, given the frequent application in these types of "covenants" and "gentlemen's agreements" that predict the repurchase of a portion of the instruments issued at the end of the chain.[30]

In this context, certain financial practices appear to be interesting. I am referring, firstly, to the option of outsourcing the process of securitization to the asset managers, entrusting them with the task of setting up an (alternative or not) investment fund (to hold the credits, in exchange of its shares, to be placed in the market in exchange of money).[31] In the European system, after the adoption of Directives 2009/65/EC and 2011/61/EU, this practice is under the prescription of an advanced legal framework, such as to require preventive indication of the policies of investment and disinvestment, to which relates the ability to prevent the negative effects of a generalized default of the assets held by the fund (through capital insurance mechanisms).

From another perspective, it should be noted that the securitization in question can also be done in "synthetic modalities," proceeding only with the transfer of credit risk, without changing the original underlying positions. In this case, there is not a sale of asset (made by the originators to the SPV), but the entering (of both) into a "credit default swap," which will be the asset used to back up the ABCPs, ABSs, or CDOs.[32] It is clear that, in this case, there will be the negotiation of financial derivatives suitable for use in the following stages of the shadow credit intermediation process (up to the final CDOs).

Therefore, the arranger shall take into account the legal content of these derivative contracts, which is based upon, on the one hand, the cash flows generated by the aforementioned loans and credits (recorded in the assets of the bank) and, on the other hand, the cash flows required by the SPV to satisfy its obligations. In this case, the reliability of the latter is not given by measuring the assets of the vehicle, but should be calculated with respect to the contents of the derivatives.

In any case, the securitization techniques remain aimed at transforming the timing of the underlying obligations of the shadow banking system. According to the *prices* of the financial market, the arrangers try to reduce the maturity of the operation, by issuing securities with a timing shorter than the one provided in the loans' contracts, to improve their profits. Furthermore, those techniques involve the use of debt, in order to achieve higher returns on equity (when interest rates are low) due to the effective use of financial leverage. Obviously, the costs of these efforts are at the basis of the differences in the risk profiles of the loans and those of the CDOs.

Once again, one can feel the limits of a supervisory framework based on self-regulation, unable to achieve the appropriate levels of transparency, and then to ensure adequate knowledge of the underlying assets or of the methods used to achieve those effects of credit transformation.[33]

Consequently, the criticisms that, with regard to the sector of securitization, give it the classification of "absent wealth market" cannot be refuted.[34] Moreover, in the shadow banking system, the lack of transparency makes it more difficult to understand the real value added by these operations (to the global wealth).[35]

Concluding on this point, it should be highlighted that the current techniques of securitization do not appear to be able to promote steps towards a safe and sound transformation of the loans granted to the real economy into CDOs. These operations seem to suffer the contradictions of an activity carried out by focusing on the volumes and profitability, and not on the risks. This is the result of the lack of "automatic incentives" able to promote the prudent composition of the assets; hence, one finds the space (and the reasons) for a regulatory intervention that attempts to make the configuration of the shadow banking operations more and more balanced (in terms of safety of the credit enhancement, maturity transformation, and leveraging).[36]

5.6 Ratings (in the shadows)

Now the consequences of the lack of transparency and of the practice to use the "ratings" in the shadow credit intermediation process can be analyzed. The ultimate goal of this part of the research is the understanding of these assessments and their impact on the efficient configuration of the shadow banking operations. Even in the shadow banking system, the credit ratings are the main tool able to provide specific information, granted by the reputation and the responsibility of the agencies that produced the data.

So far it has been considered that, in the shadows, the offering of securities to (private or institutional) investors is the final stage of a sequence of complex operations (aimed at the conversion of credit maturity and liquidity) that—as it has been pointed out—starts with the securitization of the loans granted to the real economy.[37] Obviously, this phase makes possible the transformation of the credit that is able to circumscribe (within certain limits) the possibility of the relationship underlying the securitization affecting the securities issued. Moreover, it is highlighted that this offering marks the transition from the bilateral context of (the sale of asset to) the SPVs to the multilateral one of the market in which are *placed* the ABCP, ABS, or CDO representing the securitized assets.

Consequently, I shall now consider that the price setting of financial instruments follows the traditional market mechanisms related to the fair demand–supply matching. So, the liquidity of the negotiations, the transparency of the information, and the characteristics of the demand determine the fair market value of these securities. It will also be taken into account that each operation of re-securitization introduces new *inputs* able to influence the fair value and the market prices of the following financial instruments.

A similar effect is produced by the application of the pooling technique when the transparency lacks. In particular, there is the possibility that, in each stage of the shadow credit intermediation process, the offerings of financial instruments are subscribed by more than one vehicle; so, the initial operation of the securitization and the following re-securitizations are resolved in the storage of the relevant securities in more then one SPV (constituted for this purpose). Obviously, each vehicle can buy securities offered by different issuers; therefore, there is a dispersion of the instruments issued (between multiple SPVs), and a mingling (in any vehicle) of assets from more securitizations. Therefore, any shadow banking operation can present different (in sourcing and content) elements, combined and re-combined.

Following these considerations, one can hypothesize that these market dynamics influence the prices and then reduce the importance of the characteristics of the loans granted to the real economy (because there cannot be the conditions of transparency necessary to evaluate the original assets).[38]

I shall now focus on the assignment of a "rating" to each of the ABCP, ABS, and CDO, and on the possibility that it helps the placement of these securities. It is apparent that certain investors can only subscribe to financial instruments holding a rating, and that it can be assumed that this rating will have an obvious multiplier effect on the demand (for these securities) and, therefore, on increase in market prices.[39]

In other words, it seems possible to consider that, in the case of trading in free trading venues, the rating becomes one of the drivers of demand. Moreover, a lack of evaluation (or its rejection) may preclude the marketability of the instruments in question. In this case, the intervention of the specialized credit rating agencies (CRAs) does not constitute an accessory, but it is itself an essential step in the shadow credit intermediation process.

There is no doubt that the rating supports the "more complex and advanced securitization operations" and that the same plays a fundamental role in the correct evaluation of the instruments resulting from the application of financial engineering techniques by creative arrangers. Not surprisingly, the crisis has already shown that a lot of prices were more sensitive to the range of alphanumeric codification of raters than to the actual performance of the underlying assets.[40]

This suggests that, in the shadow banking system, the ratings—as well as qualifying as "public good" in the manner indicated by careful legal analysis[41]—fulfill an information function (with regard to the quality of the securities offered in the market). Obviously, this implies specific (positive and negative) externalities, including the risk of an overestimation of the quality of the instruments issued in the intermediate stages (e.g. ABCP and ABS). It is easy to understand how the ratings can amplify (or reduce) the relevant values and, thus, lead to an exponential growth (or decrease) in volumes and prices of the final instruments (i.e. the CDOs), obviously in pro-cyclical mode.[42]

To summarize: the regulator shall take into account that, when the transparency lacks, the ratings not only allow preliminary identification of the quality of the securities, but also affect their demand and, therefore, the process of pricing. And this seems to be true from the initial securitization, through the subsequent re-securitizations, until the issuance of CDOs (offered to the final investors). This leads to the identification of a need for the regulation of this direct effect on the value (that can be also recorded in the financial statement relating to the financial instruments in question).

In light of the foregoing, we can understand the importance of the ratings, since their role appears to influence both the accounting values and the regular course of the entire shadow credit intermediation process. Obviously, to this *attitude*, corresponds a specific market power of the CRAs that formulate the aforementioned judgments.[43] In addition, the market power of these CRAs can be highlighted, which shall be limited by the public intervention in order to ensure a fair distribution of the wealth produced by the shadow banking system (i.e. resolution is not to the detriment of other service providers and investors).

Nowadays, only the arrangers and the market managers can handle these variables. Thus, these operators shall control the risk of default of the shadow banking operations under their scope. Bear in mind that such a variable should suggest to the public supervisors an intervention aimed at regulating the design phase of the shadow credit intermediation process, in ways that involve these agencies (whose judgment should facilitate the selection of paths more suitable to linking the subjects in deficit with those in surplus).

Indeed, the involvement of raters in the construction of the shadow banking operations (together with the arrangers) could undermine the exercise of "independent judgment" required to ensure proper assessment of the quality of the securities in question.[44] One cannot fail to mention, however, that the recognition of a specific role for the CRAs (in this matter) could lead to other risks, given the known conflict of interests in which, frequently, they operate (against the position of shareholders and stakeholders of the companies subject to their evaluations).

In this case, in fact, the role of agencies evolves from simple "fact checker" (of the creditworthiness of the issuer) to "partner" of the arrangers of a shadow credit intermediation process. And this is not a role that can be reduced to the production and certification of safe and liquid securities through the financialization of the loans to the real economy. On the contrary, this is a business that, usually, ends up contrasting with the regulatory frameworks of the countries in which agencies offer their service.[45]

Hence, there is a need to achieve a system in which new rules are intended to monitor the correct delivery of the ratings related to securities issued as a result of any shadow banking operation. In the latter, in fact, we have to cope with the criticality of the relationship between ratings and creative financing mechanisms.[46]

There is no doubt that such a requirement is only partially reflected in the forms of supervision indicated in the EU Regulation 1060/2009. Similarly, this also has to be said for the new provisions of the Directive 2013/14/EU (in order to counter the over-reliance on credit ratings). This applies, in particular, to the limited value of the disclosure obligations set out therein, which—in the opinion of European bodies—should be able to ensure that the issuance of a credit rating is not affected by any conflict of interest, existing or potential, or business relationship involving the agency issuing the rating (art. 6, EU Regulation 1060/09).

Moreover, the limits of such an intervention are clear when accounting for its purpose to overcome the operational practices that rely excessively on credit ratings for the conduct of their investments in debt instruments, often omitting to evaluate themselves the creditworthiness of the issuers of such instruments. Hence, there is a risk of failure to achieve the goal of reducing the excessive reliance on credit rating aside from the EPAP, the UCITS, and AIF (Recital no. 2 and 5, Directive 2013/14/EU).

It should also be considered that (at times) the industry and the investors do not know the procedures of rating elaboration (and of financial risk estimation), nor the inputs that these CRAs use. This has the obvious consequence of making difficult, for others, any check on the correctness of the analysis performed on a specific shadow operation. By waiting until the final implementation of the new rules on transparency and disclosure (together with the application of the new obligations imposed by the Directive 2013/14/EU) it will be possible to produce positive externalities for the shadow banking system, avoiding undue interactions between rating and reference values (and, therefore, reducing the pro-cyclical effects that are determined by the first).

In addition, a further element of reflection must be considered, due to the recent increase in the presence of "naive investors" within the most common trading venues. I am referring, in particular, to those who, being deprived of financial literacy and evaluative capacities are not able to measure the risk related to the offering of complex financial instruments and, therefore, make their choices based on simplified information (e.g. those which are of rating).[47] It is clear that, in the presence of favorable judgments (for a given operation), these subjects are able to increase their demand towards the instruments for which they were pronounced agencies.[48] Thus, the offer of securities to subjects inexperienced in over-rated securities determines the payment of a higher price than that properly attributable to the underlying assets, with obvious overestimation (amount) of the intermediate securities, first, and CDOs, second.

There exists, therefore, a reality in which the book values can be altered by the combination of the behavior of naive investors and redundant complexity of securitization. Hence, the risk that the CRAs are involved with operations of dubious transparency and professional integrity in order to maximize the profits of those who organize the production cycle of CDOs.

It is important to realize that this problem has a wide influence on the financial market. Therefore, it is useful to point out that a first line of defense against this risk has been applied to the obligation of "double credit rating of structured finance instruments" (i.e. double evaluation made by two different rating agencies), recently predicted by art. 8c of EC Regulation no. 1060/2009, as emended by the EU Regulation no. 462/2013.[49] This is a regulatory option that relies on the correctness of the ratings, thanks to the competition between agencies (which should compete in terms of quality and thus improve the reliability of the ratings rendered).

Undoubtedly, under these circumstances, there is the danger that CRAs will not be able to effectively measure the credit risk of these securities (as compared to market valuations influenced by the aforementioned behavior of naive investors). It is, therefore, necessary for an intervention that will incorporate new accounting rules of the assets used in the securitization transactions. That is, given that these rules must be able to ensure the compatibility of the evaluation criteria with the sequential structure of the shadow credit intermediation process and, therefore, avoid pro-cyclical effects in the production of CDOs.

6
Non-Standard Operations in the Shadow Banking System

This chapter will consider the problems related to the issuing of ABCPs, ABSs, and CDOs. Before that, I clarify that the use of securities lending and borrowing or other agreements is connected with the process velocity (in circulating the money), then must be monitored by the monetary authorities.

The chapter goes on to analyze both the (general) use of derivatives and the (specific) management of national states' sovereign debt in the shadow banking system.

I move from the hypothesis that any SPV can take advantage of the opportunities offered by the market structure of the shadow banking system, in order to verify whether this behavior produces effects that in turn affect the financial markets and the banking industry.

6.1 The use of securities lending and borrowing, and repurchase agreements

It is necessary to develop in more detail the analysis of the shadow credit intermediation processes that are not based on standardized agreements (made by international associations of market participants or other bodies). It is apparent that the most important problems can be substantially the same as those that can be been discussed when dealing with the standard operations used in credit transformation (at the basis of loans to real economies and the securities received by the cash pooling). However, I believe that the use of standardized agreements can lead to the set up of certain pieces of "soft law," which are able to drive the market to the maximization of its qualities.

In dealing with non-standard operations (NSOs), this chapter will show the shadow transactions that join the securitization of assets with "securities lending and repo or, more generally, collateralized borrowing."[1] These techniques allow the transfer, for a specified period of time, of a set of assets and, at economic level, can help an SPV to take advantage of the opportunities offered by certain market practices developed in recent times (also to stem the prohibitions of short selling[2]).

In this regard, it should be kept in mind that the aforementioned assets are able to *enhance* the intermediate stages of the shadow banking intermediation process, due to the fact that "the crucial economic function of market-making entails intermediaries going short, and so needing to cover their position by borrowing securities."[3]

It should also be noted that the activity of "securities lending and borrowing" allows a vehicle to lend or borrow securities, towards the payment of a fee. This is a case that can be referred to the contractual framework of the Global Master Securities Lending Agreement prepared by ISLA (International Securities Lending Association). The scope of the contract is as follows: "from time to time the Parties...may enter into transactions in which one party (Lender) will transfer to the other (Borrower) securities and financial instruments (Securities) against the transfer of Collateral...with a simultaneous agreement by Borrower to transfer to Lender Securities equivalent to such Securities on a fixed date or on demand against the transfer to Borrower by Lender of assets equivalent to such Collateral."[4]

For the purposes of this study, rather than focusing on traditional issues related to the distribution of dividends and the exercise of administrative rights, an SPV should be analyzed by focusing on the dynamics of *delivery*, *collateralization*, and the provision of *warranties*, as well as on the rules responsible for the management of important *consequences of a default event*.[5]

It cannot be doubted that in any case the shadow credit intermediation process will benefit from the commitment of the lender, but it is clear that this commitment is alined to the obligation to deliver certain securities to the borrower (or deliver such securities in accordance with an agreement and certain terms set in the relevant loan).[6] In contrast, that of the borrower shall require him or her to undertake (to deliver to or deposit with the lender, in accordance with the lender's instructions) a specific collateral (simultaneously with delivery of the securities to which the loan relates and in any event no later than close of business on the settlement date). This commitment provides mutual guarantees regarding the fact that the first is absolutely entitled to pass full legal and beneficial ownership of all securities provided by it hereunder to the borrower free from all liens, charges, and encumbrances. The second is absolutely entitled to pass full legal and beneficial ownership of all collateral to the lender free from all liens, charges, and encumbrances.[7]

In this context, the contractual provision related to the management of the default of one of the contracting parties becomes of particular interest. In the event that an SPV becomes insolvent, in fact, the ISLA contractual regulation requires the early settlement of mutual obligations that will have to be parameterized to a certain "default market value."[8] This has obvious effects on the "equal treatment of creditors" and on the order of priority for the repayment of other creditors.

From another perspective, take into consideration the repurchase agreements (repos), which—depending on the contractual model Global Master Repurchase Agreement chosen (among the ones proposed by the ICMA and SIFMA)—allow the sale of securities in exchange for money, under the agreement of being able to successively repurchase the same securities at a given date and at a fixed price.[9]

From a legal perspective, the difference between the two operations (securities lending and repos) appears in the relationship that is established between the parties of transaction and the financial instruments in question, as only the repos are able to produce the transfer of the property right. And this, of course, occurs in a context in which the financial instruments are, usually, fungible.

The chapter shall also consider the ways in which such transactions interact with the shadow credit intermediation process. There can be no doubt that a person who owns a portfolio may design a shadow banking operation through access to securities lending and repo markets. In its simplest form, in fact, the (temporary) assignment of the instrument in question (sometimes resulting from a securitization) occurs after the payment of a sum that may then be used to grant loans (to be securitized) or to gain other instruments (even for ABCP, to be transformed into ABSs or CDOs). Hence, one of the effects of leverage and maturity mismatch that gives content to the phenomenon under observation. Of course, the freedom of the shadow banking system allows more complex interactions, in which the party who enters into contracts of securities lending and repo can intervene in partial mode (as a co-financier of the transaction), in one or more stages (acquiring part of the securities issued or financing the vehicle by debt) or in collateral modalities (lending securities in the presence of specific safeguards).

Moreover, it must be taken into account that it is the Bank of England itself who states that "the first lender of securities might lend against securities collateral and do no more; that is relatively common in

European markets. But the entity that has borrowed those securities could themselves repo out the borrowed securities for cash, and employ the cash in a lending or credit-asset business."[10]

In brief, it can be said that NSOs can be inserted into each stage of the shadow process. Hence, the danger that the risks of the instruments transferred (under a lending transaction or a repurchase agreement) will amplify those properties of the shadow banking system and, with them, (both risks) interact—in case of market turmoil—within instant mechanisms of protection (according to which secured lenders try to realize their collateral instantly upon default).[11]

From another perspective, it should be noted that some studies have shown how the use of NSOs increases the levels of profitability of shadow banking and, therefore, influences the conditions of the offering of loans to the real economy.[12] Empirical evidence for this is alleged, according to which "contractions of broker–dealer balance sheets have tended to precede declines in real economic growth, even before the current turmoil."[13]

Therefore, public intervention of supervision is necessary, in order to limit the risk that the secured lending or the repo are exposed to the fallacy of the "lump of liquidity." Hence, there is a need for rules that, on the one hand, reduce the freedom of operators in the assumption of the risk in question and, on the other hand, entrust monetary authorities with the task of managing the events in which "when haircuts rise, all balance sheets shrink in unison, resulting in a generalized decline in the willingness to lend."[14]

It can also be said that, at least in part, these aspects have been covered by Directive 2009/44/EC, which has innovated the European rules on "settlement finality" and "financial collateral arrangements." In addition, the ECB seems geared toward the consolidation of information on those operations, promoting the development of an EU database, "because market interconnectedness knows no borders, a global alignment in terms of repo market transparency is desirable (in results, not necessarily in solutions)."[15]

In this context, it cannot pass unnoticed that the key drivers of the securities lending and repo markets are aligned to those of the shadow banking system. Therefore, the review of the regulatory system must ensure the stability of both, avoiding different but related problems. This applies, in particular, to the negative consequences of the lack of transparency, to which is linked the interconnectedness through valuation, haircuts, and, therefore, the pro-cyclicality of the leverage system. Other issues are related to the re-use of collaterals, and more generally to insufficient rigor in collateral management and valuation.[16] Thus, the need to ensure that investment firms and fund managers can take advantage of NSOs to circumvent prudential regulations, working on tools that they could use in a direct way.[17]

The operations that give content to the "tri-party repo market," where the facilities provided by an intermediary bank allow the parties to use a clearing service safely, are also significant.[18] However, the application of the "daily unwind process" allows financial instruments to come back to the availability of the transferor during the opening hours of the market; and the same are then returned at the closing of negotiations on a daily basis. This means that the hazards associated with these operations occur twice a day, and then the exposure intra-day is particularly risky, and difficult to manage even by the eventual application of the current prudential supervision.

That said, the practice of "re-hypothecation of client assets" should be considered. The application of this practice improves judgments on the quality of the instruments issued at the end of the shadow credit intermediation process. Not infrequently, the arrangers hold financial instruments belonging to third parties. These instruments, however, are not always stored in a "segregated client account," but are used to increase the quality of the shadow banking operations, including the assets of the SPVs that proceed the offering of ABCPs, ABSs, or CDOs.[19]

This results in a situation in which it is possible to use the free availability of other people's resources as shadow banking entities. The latter can then use them to fund their activities, leading to a confusion of assets and, moreover, a condition of opacity incompatible with the regulatory criteria of the financial market.[20]

All in all, it can be said that—once again—there is a need for increasing the transparency of the shadow banking system, even when it relies on standard agreements. I feel that, at least in part, there is a need for shadow banking operations to be linked to "trade repositories" that are able to carry out the business of publishing the operations' data (at least at the aggregate level).

6.2 The offering of the asset-backed commercial paper, asset-backed securities and collateralized debt obligations

In attributing a peculiar relevance to the offering of the financial instruments, I have expressly assumed that the quality of the ABCP, ABS, or CDO depends on the ability to introduce solvable "loans, leases, and mortgages" in the shadow credit intermediation process, followed by implementation of safe credit enhancement operations. Further reflections show, I think, that the shadow operations should improve—at every stage of production—the quality of the securities placed for the collection of resources available in the capital market.

Despite this, the offerings of financial instruments take place in a globalized market for interconnection, but not for adjustment. So, the arranger of the shadow banking operation may want to access a particular (regulated or not) trading venue not only on the basis of the factual need to go looking for the centers where the demand—of liquid and sure financial instruments—is not yet saturated, but also the possibility of anchoring certain stages of the production cycle centers offering a competitive regulatory framework.[21]

Therefore, the focus here will be on the possibility that the establishment of any SPV—and, hence, the offering—is subject to the rules of a market that attracts the industry for the freedom and simplicity of its structure (also at the level of self-regulation).[22] In the selection of the target market, in fact, there is a situation of *full economic antagonism*, which interacts in a systemic competition that can induce a state to increase its ability to attract such investments through the configuration of a particularly *flexible* national legal system.

Following this approach, it can be assumed that the choice of the type of securities to be offered shall be generally oriented towards asset backing (i.e. ABSs or ABCPs) or collateralizing (CDOs), depending on the features of the demand. In this case, in addition to the economic conditions of the financial market (and, therefore, the individual preferences of those who will demand this type of instrument), the arranger shall consider also the regulatory framework and, therefore, the freedom that the law recognizes the configuration of individuals' rights that can be embedded in such financial instruments.

Consequently, the dynamics—of transformation of the credit that give content to the shadow banking system—seek the best environmental conditions, both economic and juridical. This is the result of the "regulatory competition policies" adopted by certain countries to attract investment and financial firms. These policies allow them to take advantage of asymmetric regulations, and thus facilitate the implementation of cross-border transactions that are not always transparent (but functional in their objective to reduce the burden that would have been observed in a traditional social-democratic system).[23]

It is clear, then, that the *political foundation* of the instance aims to increase the level of transparency and security of the shadow credit intermediation process (and, more generally, of any intermediate stages of any form of market-based financing). National governments seem to be oriented towards preventing resolution of the competitive drive in question with an increase in the overall risk of shadow operations. In the requirement for public supervision, therefore, the intention is not only to ensure the correct application of known legal institutions—provided by the law—to transform the credits deriving from the financing of the real economy (and in particular of the discipline of securitizations). It is also aimed at—most importantly—identifying the devices that can ensure the overall stability of the capital market, and thus prevent a crisis of the phenomenon under observation being extended to undermine the welfare levels achieved by economic development.[24]

There is no doubt, in fact, that the issuing operations of ABCP, ABS, and CDO result in an offer entirely destined for the market, free of those subjective reserves provided to limit the audience of investors (and, consequently, to avoid the negative consequences of any possible defaults passing to the detriment of only specific subjects). It is, therefore, possible that, even in the middle stages of a shadow process, the financial instruments are subscribed in part by the persons who will carry out the next operation of the transformation of credit, even though the rest may be acquired by third parties (or remain in circulation on the market, as a floating).

In evaluating this opening to the market of the shadow credit intermediation process, the limits of the presence are felt—already at an intermediate stage—of the fair market value accounting principle for the recording of securities ABCPs, ABSs, and CDOs. This, then, is the point of initiation of one of the components of the pro-cyclicality; and indeed, the increase in market value of intermediate assets (including loans and CDOs) interacts with the levels of leverage and, therefore, amplifies the value of the assets underlying the final offerings and reference volumes.

At the same time, the prospective of an interventional action by the supervision authorities and national regulators should promote a reorganization of the shadow banking system, where the will of the industry—to arrange the most transparent and secure mode—could allow a rapid evolution of paradigms in order to match the level of protection that should characterize the capital markets of democratic systems.

It is in this way, in fact, that the operators themselves can offer (or, rather, suggest) to the aforesaid authorities a way to understand and monitor the phenomenon under observation. Indeed, the operator can suggest a reason to preserve this data that may indicate the positive effects of competition in the banking system, in terms of competitive pressure (and, therefore, lower prices and higher volumes, subject to increase in the supply of alternative investment products).

It is clear how the offerings lead back to the issue of bringing the shadow banking system under government oversight. Therefore, any entity should comply with the rules of the market, in order to sell its financial instruments. This indicates the shadow credit intermediation process consists of operations that go beyond the economic space that blurs the boundaries of the financial market. This consideration suggests specific questions as to whether the supervisory authorities shall continue only to monitor the SPVs issuing the final instruments as a single issuer or, rather, focus on the whole set of credit and credit transformation operations used in the shadow process.[25]

6.3 Peculiarities of derivatives

At this stage of the investigation, it shall be agreed that the shadow banking system provides a specific daisy-chain of market operations. It should also be clear that these operations may seem independent on the legal-formal level; they are economically connected in relation to the performance of capital flows. It goes without saying that, in any case, it is possible to use derivatives in order to increase the efficiency of the process as a whole and, therefore, to improve the effects of the credit transformation that is set to be made.

Once again, take into consideration the design of the shadow process (and, in particular, the selection of the types of operations that transform the granting of loans to the real economy in safe and liquid instruments, to be offered to the market).[26] I shall refer to the flexibility of the clauses that define the profile of the performances at the basis of any derivative contract (and, therefore, the possibility to adapt its content to the needs of each stage of the shadow credit intermediation process). It is apparent, then, that the arrangers can, alternatively, purchase derivatives in the market (regulated or not) or proceed on to bilateral negotiations (with a private party) at any stage of the shadow process, and using any of the various SPVs involved in the transaction.[27] However, market operations must necessarily comply with the rules of trading provided for the relevant venues (and shall also qualify for the application of safeguards to ensure the efficient movement of capitals), while bilateral operations are subject only to the contractual regulation, which must ensure a substantial symmetry of the costs and of the benefits.

Of course, in any case, only the expertise of the contracting parties can ensure the reliability of the forecasts and scenarios, and only this can avoid a mutual influence setting the stage for increased levels of systemic risk.

This suggests that, in the shadow banking system, there is a specific interest (of the arrangers) in the quality of the derivatives (placed *a latere* of the production cycle in question). Hence, there is a need to ensure the fairness of the algorithms chosen to give content to the derivatives (in a manner consistent with the will of the parties and, therefore, respondent to their expectations). Therefore, in the absence of any supervision (or, at least, any monitoring system), attention must be focused on the existing market practices and on the responsibility of the shadow banking entities that choose to include certain derivatives in the operational phases of their competence.

That said, it should be noted that the derivatives shall satisfy the needs of the credit transformation (with regard to duration, maturity, and liquidity enhancements). Hence, the arrangers shall select the ones whose value depends on the performance of the underlying financial instruments and their (direct or indirect) relevance to the shadow banking operation. In this regard, the importance of "forward contracts" should be taken into consideration, where the binding commitment to buy or sell an asset (at a predetermined price and date) is the legal instrument necessary to take short or long positions that may allow, in the intermediate stages, a response to the needs of the coverage, exchange, or liquidity. Similarly, it has to be said for the "futures," however, that there is the application of standardized contract formulas, which allows its trading on a regulated market and, therefore, the applicability of the protections provided by the system (in terms of price transparency and risk mitigation).

Again, even the role of the "options"—whether put or call—is preordained to the improvement of the production process, given the ability of a subject to pre-determine the conditions of sale or purchase of financial instruments (i.e. ABCPs, ABSs, or CDOs). In other words, an option may place the entities in a condition of relative safety compared to the market trends (of course, subject to acceptance of its purchase cost).

More complex is, however, the use of "swaps," because of their versatility that can lead to other effects in comparison to the exchange of cash flows (calculated on the basis of different parameters).[28] In the shadow banking system, in fact, the conclusion of such a contract may allow both to apply the technique of synthetic securitization (through the negotiation of credit default swaps), and to increase the levels of leverage (including in this case a *loan element*). From the initial securitization of loans, such a derivative will be able to consider both the portion of interest and that of capital, where an "amortizing swap" is a functional transformation that gives content to the phenomenon under observation. This is irrespective of the possibility of introducing in the derivative's regulation an upfront clause (also embedded) that—in each stage—can cope with sudden liquidity needs.

It is important to consider the economic literature that has highlighted the awareness of traditional banks for derivatives used in the shadow banking system. This is an awareness that results in the goal to better manage their assets, employ at high rates the liquidity excesses, and, ultimately, improve their process of "borrow-short-and-lend-long."[29] In such cases, the shadow credit intermediation process must involve individuals who act as hedgers, mitigating the risks related to changes in interest rates and in the volumes of demand. It cannot, however, be ruled out that the shadow banking entities also act as arbitrageurs and speculators, betting on the future of the market or brokering complementary positions for the sole purpose of achieving an immediate profit.[30]

With regard to the selection of the contents of the derivative (and its compatibility with the transaction being entered into), it should be noted that a correct option in this sense is closely related to the degree of expertise of the arrangers.[31] The result is that individual professionalism and fairness are the only safeguards against opportunistic behavior and moral hazard that may be detrimental to a single transaction or the entire intermediation process. This is an evidently insufficient safeguard to meet the needs of stability and security that characterize the capital market and the agents who access it.

It must also be said that the possibility of systemic risks derives from the forms of centralization of derivatives and, therefore, from the regulatory choice to entrust the task of managing and ensuring the settlement of the transactions in question to spcialized companies. Thus, it will be these companies that identify the standards required to access their services; hence, the need for adequate guarantee mechanisms to prevent the crisis of a shadow banking entity placing the whole system of compensation in difficulty. Basically, if the doubt should consider that the SPV cannot offer guarantees more than the assets they use (to give content to the stage of the shadow credit intermediation process within their competence), then, there is a real risk that the mechanisms based on the use of collaterals can—put to the test—prove themselves insufficient to ensure a supply alternative in the event of default.[32]

Finally, the role of the International Swaps and Derivatives Association (ISDA) in the highlighted context must be emphasized. In promoting the adoption of "legal standards" in the construction of reliable operations in derivatives, ISDA has directed financial firms and brokers to practice consistent with the behavioral criteria required by the supervisory authorities. It is clear that, in the shadows, the derivatives—agreed in accordance with standards generally accepted by the financial community—reduce legal and counterparty risks, which occur in such cases.

It can be supposed that the standardization of contractual texts that will be able to facilitate the diffusion of derivatives requiring the intervention of a central counterparty shall reduce the risks associated with derivatives that—as evidenced by the Rules EMIR no. 648 of 2012—are managed using alternative risk mitigation techniques. Moreover, according to ISDA, the "EU proposal on Structural Reform of the EU Banking Sector," in reducing the activities of the "core credit institutions," results in "encouraging clients to interact with the shadow banking sector—rather than the more regulated and less volatile (in terms of it being a source of funding) banking sector."[33]

Thus, it is possible to anticipate an early conclusion with regard to the regulatory intervention *in subiecta materia*. More specifically, I refer to the ability to direct the use of derivative contracts towards a balanced economic structure of the shadow banking operations (i.e. such as to protect the capital invested by wicked desires of individual profit maximization and, therefore, from opportunistic attacks aimed at distorting the process of credit intermediation in question).

However, it should be noted that the transactions in derivatives, when traded on regulated markets, are another factor of interconnection between the shadow banking system and areas of supervision. This is not only because of the direct relationship that develops between the shadow banking entities and its counterparts in negotiations, but also the effects that the default of the first might have on the central counterparty to which the settlements are remitted.

More generally, it can be said that the critical issues mentioned just now are associated with the absence of legal limits on risk taking, leverage, and ultimately the exasperated search for innovative techniques to be applied in order to earn an individual and immediate profit, without bothering to check that all of this is compatible with the needs of sustainable development at the basis of contemporary economies.

6.4 The shadow banking operations of sovereign states

Two particular aspects of the shadow banking system must be investiagated. The first concerns the operation of the sovereign states out of the regulated markets, and their subsequent use of derivatives aimed at financing public policies through direct access to alternative sources of funding. This is related both to the activity that some countries have drawn up to cope with the turmoil that in the last few years has hit most of the euro zone and the difficulties in refinancing sovereign debt of certain member states of the EMU. The second aspect, which will be studied in the last paragraph, concerns the shadow operations of the banks, given the danger that these are the main channels for transferring the risks from one system to the other. Both these aspects are, in fact, related to the relevance of the operations that generate interconnections between the shadow and the regulated market (where sovereign states and banks should operate only in the latter, for obvious safety reasons).

The issuance of sovereign debt—and the related open market operations—identifies only one of the most known and important forms of market-based financing entered into by sovereign states. In this regard, it should be noted that, not infrequently, derivatives complete the range of public interventions in the capital market. They are not so much authoritative interventions as activities related to the desire to achieve a given economic result through the activation of bilateral or market relationships (and, therefore, by the conclusion of certain transactions aimed at acquiring capital or derivatives to ensure certain equilibrium). These activities are in line with the empirical observations that evidenced the use of derivatives by sovereign borrowers; and in particular have shown that "swaps are used both to increase the liquidity of long-term government bonds and for speculation." This is aside from the fact that "some sovereign borrowers have also used derivatives to 'window dress' their public accounts for the purpose of disguising budget deficits."[34]

Moreover, it is known that the position of any government that has access to capital markets to cope with uncontrollable expenses (or, rather, excessive deficits) and to refinance enormous debts is weak. Thus, the shadow operations concluded by public administrations are clearly intended to reflect the gap between member states within the euro zone that are aligned with the monetary parameters and the countries that have not taken that course of action (i.e. central European *versus* Mediterranean).[35]

In this context, it must be determined whether the "convergences" implemented in view of the EMU have given rise to an actual homogenization within the eurosystem or whether these are not sufficient to ensure the establishment of a sustainable common development strategy. In other words, there is evidence of the risks associated with the choice of certain member states to keep a large degree of freedom and, therefore, to operate in the absence of a safety net designed to contain the borrowing policies.

Therefore, doubt arises about the closer supervision of a portion of the internal market, and, that is, on the public finances of the member states of the euro zone. This supervision does not appear able to prevent certain refinancing transactions of sovereign states being made within the shadow banking system (either through the issuance of public debt or the entering into derivative contracts). This doubt is only partly sedated through reference to the role played in the European Union by the ESFS, given that only the new "building blocks" of the supervision on the European banking and financial industry (and then the beginning of the current EBU and the forthcoming Capital Markets Union) would be able to align forms of public intervention in the relevant capital markets.[36]

There is, therefore, a danger that those innovations—brought in the pursuit of stability—create differences between the two European associative formulas (the European Union and the EMU), so as to empty the political contents of the first (against the failure to achieve with the second a degree of cohesion ideal to project countries that share currency towards a political union).[37]

This suggests an instance for the revision of EU governance—and, therefore, the change of both the Fiscal Compact (so-called SCG Treaty) and the European Stability Mechanism (ESM) Treaty—in order to proceed to the regulation of shadow banking operations in which a member state intervenes. This shall be consistent with the changes introduced in the internal financial market as a result of the EBU (that completes the European System of Financial Supervision).

In the meantime, the danger remains of "black-ops" designed to shift the weight of current policies onto future generations, as does the possibility to execute them in the shadow banking system.

Furthermore, consider that there are derivative contracts subscribed by sovereign states outside the markets they regulate or within the shadow banking system itself. I will not dwell on the full capacity of private law (recognized to national public administration), but on the validity of the choice to work in a system that is unrelated to (their public) financial supervision. Therefore, this analysis does not concern only the evaluation of the structural features of the transactions in question (arising from the legal form of contracts), but the solution to the problem affecting the cost of the choice to act in the shadows, outside the transparency rules (and, therefore, in an opaque context).

It must be kept in mind that the relationship between the administrative power (on sovereign management and debt) and the shadow market has been put to the test by some of the transformations that took place in Europe after the end of the "short century." On this point, it should be noted that the interrelationship between public and private (or, rather, between power and market) must follow: (i) (financial) liberalization and (European) harmonization; (ii) stabilization (of the economy) and accountability (of public policies); and (iii) competition (including in public services) and negotiated use of the power (by the public administration).[38]

I am talking about changes that reflect the evolution of the economy and, in particular, the evolutions in the relationships between politics and administration in the financial markets. The financial turmoil of the third millennium, in compromising the basic economic determinants, has highlighted the failure of self-regulation of markets and, at the same time, the ambivalence of the role of sovereign states. If, at first, the crisis had put the state (interpreter of the role of "savior of last resort" of the economy) at the center, the same has also highlighted the difficulties specific to certain European countries in continuing to refinance their debt (and, therefore, to limit the operations of public finances within the strict limits imposed by the economic and monetary union).[39]

In other words, the recent turmoil in the sovereign debt market has shown an abuse of innovative finance techniques. Consider that this abuse was made possible by the conditions of opacity in which the competent authorities had undertaken the management of public finance. This applies, particularly, in the absence—in the regulation of the European Union—of the obligation to represent (i) the real economic needs of the country; (ii) the uses to which is destined the collection of resources on the market; and (iii) real time correlation between the first and the second (i.e. the demand and collection). Clearly, it is necessary to stress the limits of operations over the counter of the states, in order to understand the risks associated with a regulatory framework that—even after the recent innovations of the European law—does not ensure high levels of transparency, nor the overcoming of the complex web of interdependence which—in relation to the crisis—has jeopardized the stability of the economy and finance.

There are many concerns related to the fact that the apparatus and the exponents of public administration, for getting in touch with the banks and financial intermediaries, end up operating freely in the shadow banking system.[40] It goes without saying that the option for bilateral negotiation precludes—at least partially—the possibility to take advantage of opportunities that characterize the trading systems organized around a central counterparty.[41] This, of course, also leads to anti-competitive effects, given that—in the absence of a multilateral market—the levels of transparency of information are reduced and the transaction costs increased.[42]

We cannot cast doubt on the fact that the current rules allow member states to address the shadows for purposes *other* than the mere reduction in financing costs. In fact, this is reinforced by the fact that "operability outside of regulated markets" takes place in a context of lack of transparency and total absence of a central counterparty; hence, the full assumption of liquidity and regulation risks, with obvious negative effects on the disclosure of such transactions in respect of the European control organisms.[43] Moreover, consideration must be given to the uncertainties that are related to the difficulty of containing the risk taking on the part of the public administration that subscribe to such contracts, with respect to the negotiating counterpart, the object of the derivative, and the cost-effectiveness of the operation.[44]

Concluding on this point, it can be said that certain shadow operations expose the public finances to the business conditions unsuitable for any public administration, given that they are concluded outside of the supervision and in the absence of a specific accountability and disclosure policy (which should ensure the efficient pursuit of the general interest).

6.5 The shadow banking operations of credit institutions

As was anticipated in the previous chapters, specific risks arise when banks provide services or credit to the shadow banking entities, and in particular when banks finance the leverage of the SPVs. These risks are linked to bilateral transactions that, in general, are based on special agreements (and, in some cases, the subscription of derivative contracts).

Among the financing operations offered by banks, this section shall consider the short-term ones, including the opening of a credit line (useable to manage cash flows associated with the purchase of ABCP and the issuance of ABS). To the elasticity of this relationship corresponds the possibility for the SPVs to have a reserve of money that can reduce the risk of the operations' illiquidity. The need to limit the costs of these operations leads, often, the vehicles to move towards other forms of financing, including on sight, to be renewed in the continuum. I refer in particular to the absorption of the offer of "hot money" from the banks (which is implemented willingly by the market, as proposed in terms of the benefit). These lines of credit (granted by banks to shadow banking entities) can produce variegated effects: the amplification of the leverage effect (and, thus, used to buy more assets than was possible with the outcome of their offerings), the coverage of the risk of cash or roll over (and, thus, cope with sudden demands for repayment or difficulties in refinancing operations), and the increase of the profitability and certainty of the shadow operation (with the possibility of getting a higher rating of CDOs).[45]

It should also be pointed out that credit institutions have the right to invest their own assets in the financial markets, with the result of determining a variability of the accounting values referring to bank assets (related to fluctuations in the prices of the securities purchased, among which may be present also CDOs). This has the obvious consequence that developments in the capital market reflect the stability of the latter.[46]

It is also important to highlight that the benefits arising from these lines of credit (and, in particular, from those set up as stand-by credit lines) may encourage the industry players toward opportunistic behavior, such as to orient the instrument in question to covering an uncertain need (as in length and depth). It is apparent that there is the possibility of a pro-cyclical use of these lines, such as to influence the determinations of the monetary policy and, therefore, to over-influence the real economy. This risk identifies a downside to the transactions in question, as it interacts with the banks' balance sheets and, therefore, becomes a source of systemic instability where the insolvency of a shadow operation extends its effects in a pernicious domino effect.

With regard to other transactions that may involve the banks in the shadow banking system, it is necessary to refer to the intervention of support (or, rather, of guarantee) and securities placement. In the first case, banks offer a service to ensure an alternative source of money, in the event of the issuer's insolvency, with the assumption of a shadow risk upon payment of a fee or premium. It is not, therefore, an operational interconnection, but the provision of a service (of a kind similar to insurance); hence, the level of the risk in question lies in the ability of the bank to price this service and, more generally, to safeguard the technical provisions in place to face it.

As for the placement of shadow financial instruments, it is easy to remove the doubt that the offerings respond to specific needs, sometimes connected by the optimization of the distribution of risks in the market.[47] It is clear, in fact, that the promotion of such securities to the banks' customers raises competition problems (between the offering of the same and the traditional forms of deposit). In this case, the exposure of the bank is lower than that provided for the provision of a guarantee for the success of an operation (and, therefore, for the achievement of minimum volumes of subscription of CDOs).

Conversely, it does not appear easy to solve the problem concerning the indirect exposure of banks to the risks of the shadow banking system. To it are opposed the establishment of the principle of concentration of trading in CCP (as per EU Regulation no. 648/2012) and the new levels of attention required by Directive 2014/36/EU and EU Regulation no. 575/2014 for the management of counterparty risk. In both cases a path is indicated to reduce the possibility that the banks find themselves operating in areas contiguous to those of the system under examination.[48]

The abovementioned doubts can be summed up in the specific question raised with regard to the operations of self-securitization (or retained securitization, which takes the form of a securitization arranged without selling to the market the financial instruments produced). Even if this kind of operation responds to the need of having securities to offer as collateral to monetary authorities for specific refinancing operations, the same must not allow any arbitrage in the application of the prudential regulation (set by Basel Accords). Hence, there is a need to limit the freedom of the operators and to include this form of securitization in the scope of the supervision, since it is not possible that the ABS (resulting from the securitization) shall be accounted (or weighted) for an accounting value different from the one of the previous loans. This is the reason why the supervisory authorities shall analyze any risk profiles of the securities (that, strictly speaking, require an accurate weight for transparency, of the assets transferred, in the calculation of capital adequacy).[49]

7
Shadow Banking Risks and Key Vulnerabilities

This chapter will review the risks of the shadow banking system and provide the background for understanding the different areas where the supervising authorities are planning to strengthen oversight and control.

This is the background of our investigation, from a regulatory perspective, of the risks arising from the organizational structure of the operations, and the governance design of the entities. Then, the focus moves on to the possibility that these risks infect capital markets. This will explain why the central banks and the supervisory authorities are going to reduce the freedom of shadows in order to avoid future events like those that happened during the financial crisis.

7.1 Areas of risk in the shadow banking system

Once again, it is important to provide (to start with) a few terms that will be defined precisely later. It should also be considered that capitals and their circulation require the operators to afford two kinds of expenses:

- the amounts paid out to the collection of money (from the investors—a sort of *factor costs*); and
- the amounts paid out for the required services (including the financial ones—a sort of *user costs*).[1]

The possibility follows of identifying two areas of risk:

- one coinciding with the patrimonial instabilities of the shadow banking entities (and associated with the quality of the assets held by them); and

- the other with the incoherencies of the shadow banking operations (and, therefore, with the unsustainable use of financial leverage or the unsound practices of credit transformation).

I shall also take into account the so-called "systemic and interconnected vulnerabilities," by which is meant the system elements that are exposed to the negative effects of "correlated horizontal shocks" (and, therefore, to the incidence of macro-systemic risk deriving from their contamination ability).[2] This goes together with the safety of the wealth–risk–revenue circuit at the basis of the democratic financial system (and, in particular, of the supervised dynamics of investment).[3]

According to the current regulation of the banking industry, the focus on risks implies the need to understand if the traditional rules suit the shadow banking system.[4] In this context, I shall take into account the centrality of the capital adequacy in the assessment of the stability of any credit institution, and the safe and sound management of its activities.[5] In particular, it shall be noted that the growing attention being paid to the risks (taken by banks) corresponds to a significant system of guarantees, preordained to balance the negative effects of management events. Then, the fact must be taken into account that the regulator has not limited itself, however, to highlight the strategic value of collaterals, but has adopted a set of rules designed to identify the elements to be calculated in order to reduce the burden of the risks on capital.

Furthermore, the shadow credit intermediation process is alternative also to the harmonized regime of risk–guarantees relations existing in the EU banking industry (which has been implemented by the member states in order to achieve greater integration and operational efficiency).[6] Besides, this legal framework has been set up to improve the banking sector's ability to absorb shocks arising from financial and economic stress, whatever the source. At the same time, this is based on the capacity of a credit institution to internalize (and absorb) the risks (in its own business).

Undoubtedly—as has been observed in the previous chapters—the same problems occur in the shadow banking system. However, it can now easily be understood that these risks may be amplified in the shadows, because of the limited knowledge of conditions that qualify the aforementioned process.[7] One should be aware that insufficient levels of transparency (and the complications in the diffusion of information) lead to conditions of "uncertainty" unable to preserve the system from matching the *competitive equilibrium* necessary to ensure maximized social welfare (and, at the same time, to prevent unjustified individual extra profits).

That said, it is helpful to recognize the relationship between uncertainty, risk, and profit: the concept of "uncertainty must be taken in a sense radically distinct from the familiar notion of risk,"[8] hence the option to report the first term only to non-measurable criticalities, and the second to the predictable and measurable negative events (so-called expected losses).

Therefore, the analysis of macro-systemic risk factors can be divided into two phases: the first dedicated to the subjects, the second to the activities. This brings, therefore, into consideration the risks related to errors in the management or control of the shadow entities, as well as the negative effects of events external to the company's organization (Table 7.1).

As shall be seen in the following chapters, these risks undermine—in diverse ways—the financial performance (and, therefore, the balance of cash flows), the capital adequacy (and, with it, the consistency of assets in comparison to the liabilities), and the economic performance (which must be in line with the expectations of those who have arranged the operations).

In addition to the regulatory framework adopted by the Basel Committee, the regular functioning of the shadow banking requires a new type of supervision, which should be able to take into account also the risks that do not arise in a bank but do in the shadow process. From this perspective, it is obvious how it is necessary to determine whether the existing supervisory system is able to manage the risks (that manifest themselves on the market in consequence to the affirmation of the abovementioned new capital intermediation techniques).

Therefore, the focus of attention is on the effects of the application of models that refer, on the one hand, to originate-to-distribute banks

Table 7.1 Areas of risk

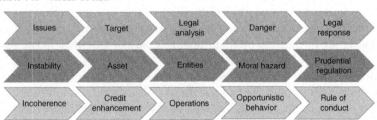

and, on the other hand, to the entities involved in the shadow credit intermediation process. Hence, the logical overcoming of the mechanisms at the basis of the Basel Accords, as these are just aimed at the assessment of the resilience of individual banking institutions to periods of stress. (Hence, their limited effectiveness at macro-prudential level and, in particular, towards the wide risks that can build up and amplify across the capital markets.)

7.2 Risk factors

In the shadow credit intermediation process, the risk of counterparty default arises as non-fulfillment (by the subjects that have been granted the funding through credit lines) or non-placement (of the securities to final investors). This means that this risk can have a bilateral character in the first case (related to the assessment of the creditworthiness of the counterparty) and a multilateral character in the second (bound to the performance of the demand made by the operators).

For the purposes of a complete assessment of this topic, it is necessary to analyze this risk in a dynamic perspective. The critical profiles of the circulation of securities, although related to real credit rights (i.e. the initial funding), qualify for the levels of leverage and for the reliability of credit transformations.[9] Hence, the failure of the ordinary prudential criteria (in relation to the objective of protecting the markets and investors), given the correlation—encountered during the subprime mortgage crisis—between the default of speculative operations and the risk of failure of the system. On this point, one must take into account the trading of the securities issued in the intermediate stages of the shadow credit intermediation process (and, therefore, a portion of the ABCPs and ABSs to be placed at the foundation of the CDOs). It will be possible to understand whether the risks resulting from the application of the principle of fair market value are present in all shadow banking operations or not. Then, the pro-cyclical effects found in the present case should also be considered—both at amplification of the creditworthiness level and of the increase of the volumes—and, therefore, the danger that a fluctuation of the economic cycle could lead to more negative results than proportional (as opposed to the actual reduction in market prices; see Table 7.2).[10]

It is clear that the abovementioned critical elements undergo a further negative effect due to the cross-border nature of the shadow banking operations, to which is related the specific threat that national legal

Table 7.2 Risk assessment

Risk factor	Event	Control measure	Prospective of regulation
Counterparty default	Non-fulfillment	Accuracy of credit scoring	Standard measure of capital adequacy
Roll over risk	Failure to repeat the operation	Analysis of market demand	Monitoring
Contamination risk	Errors in the mix of assets and in the wharehousing	Tracking of the ABCP, ABS, and CDOs	Identification of asset and legal entity
Tail risk	Remote negative events	Analysis of the operation	Additional backstops
Organizational risk	Mistakes in the intermediation process	Analysis of any stage of the shadow credit intermediation process	Monitoring
Sustainability risk	Lack of margins	Identification of the stages	Accounting controls
Runs	Unexpected reimbursement requests	Backstops and prohibitions	Regulatory constraints
Hidden leverage	Lack of transparency	Limit to debt	Regulatory constraints
Mispricing	Mistakes in the evaluation	Additional accounting control	Organizational and accounting constraints
Derivative transactions	Errors in the design of the operations	Analysis of the contracts and financial instruments	Central counterparty intervention
Governance risk	Errors in the organizational structure	Analysis of the legal entity, conflicts of interests, and potential financial pressures to the sponsors and the arrangers	Monitoring

systems are unable to allocate responsibilities to foreign entities that have produced the riskiest conducts.

From another perspective, the existence of credit transformation operations leads to consideration of the trends of the flows regarding "funding long-term assets" and their correlation with the "short-term liabilities." This analysis helps to understand how the maturity transformation is realized through market operations exposed not only to liquidity risk, but also to the so-called "roll over risk" (i.e. the possibility that, at maturity of the securities issued, the SPV fails to repeat the operation and, therefore, to refinance the stage of its shadow banking operation).[11]

Despite this, the absence of prudential limits to the indebtedness and lack of investment of personal means and own funds (in the SPVs or in other shadow banking entities) make opportunistic behavior and moral hazard possible for those who organize the operation. It is obvious that the economic performance of the subjects in question follows alternative routes, depending on their nature; because the profits are divided between the arrangers and the losses faced by investors.

Therefore, the danger is that certain operations of warehousing or pooling facilitate a "contamination effect" in the presence of default risk (relative to the subjects of the real economy to which has been granted initial funding).[12] Indeed, the default of the loans can penalize more chains of operations. In other words, the default of the instruments issued at the conclusion of each securitization may encumber the assets of several vehicles (that subscribed the emissions of ABS or ABCP). In this way another pyramidal structure can be identified, in which each stage of the process amplifies the risks (of incorrect asset valuation) related to *confusion* (in one property) of financial instruments with diverse origin (from time to time gathered in a new property). In addition, a similar effect results from the *dispersion* (in markets) of each issuance (divided between several properties, each headed by a different subjective entity).

It follows that—at first sight—this "mix of assets" (with different sources) can produce positive effects (arising from the diversification of the underlying portfolio or, rather, the fragmentation of the risk of default).[13] However, sometimes the presence of such an occurrence results in damage in the offering of CDOs (caused by an individual non-performing loan or past due). This assumption is based on the consideration that an operability of this kind increases the risk—and perhaps gives the certainty—that the SPV incurs a default (of a limited set of assets) affecting (sometimes only slightly) its operation and, therefore, produces a negative externality on the system. Hence, there is the need for introducing, against this risk, specific buffers (or similar instruments), in order to contain the effects of these losses (and, thus, to prevent the same producing shortcomings—of the cash flows or assets—able to determine the default of the vehicle).

In this case, in addition to the contamination risk, there is also the danger that the conditions of opacity facilitate misjudgments—due to the lack of information necessary to understand the trends of the market—and, therefore, an uncontrolled reaction of investors (with obvious negative effects on prices). The experience of the recent crisis has highlighted the *shortage* of those working in the shadow banking system. I refer, in particular, to the occurrence of events related to problems that can arise both within a single vehicle (due to the poor quality of the underlying loans), and in response to the malfunction of a complex series of operations (in which leverage and the maturity transformation have interacted in improper modalities).

To this lack of professionalism (and prudence) corresponds the absence of a self-organization that aims to include the negative consequences arising from the default of transactions poorly designed. That is, especially in cases where the shadow banking operations concretize longer-term credit extension based on short-term funding and leverage, free from safeguards designed to cover risks.[14] It is not surprising, therefore, that the control of the shadow banking risks is—today—one of the main challenges for the financial supervisory system. The system must respond to the need to monitor the "shadow banking...because of its size, its close links to the regulated financial sector and the systemic risk that it poses." (The construction promoted by the G20 in that field, on the other hand, will not be able to pursue the goal of "transforming shadow banking to transparent and resilient market-based financing.")[15]

7.3 Operators and internal policies of risk management

It is important to distinguish the types of events that, when they occur, can have negative effect on the economies of those who invest their capital in the shadows. And it follows that one shall take into account any single event in order to understand how to manage the shadow banking risks. As will be seen in Chapter 8, public oversight is stretched to the objective of preventing the freedom of the operators from producing risks that can undermine the smooth movement of capital, the efficient allocation of resources, and, therefore, the production mechanisms of wealth and of social welfare maximization.

Therefore, attention should be paid to the verification of cases in which a shadow banking operation does not meet the expectations of the subjects involved in it. Furthermore, this effect may be a result of risks resulting from the chosen organizational forms, as well as other external factors (which could cause a slowdown or, in the worst case, the halt of the shadow credit intermediation process). It is, therefore, of (political) general interest to ensure that those risks do not go beyond the boundaries of the shadow banking system, eventually infecting the entire capital market.

Accordingly, there is the need to enable public control on the risks in question, having regard for the interconnections between the shadow banking system and the global financial system. For this purpose, both the placement modalities in the traditional markets of the instruments ABCPs, ABSs, CDOs, and the involvement of banks, finance companies, and insurance in such transactions come into account. These controls shall have to supervise over the *official channels* in which breeds a "wave of contagion that would affect sectors subject to the highest prudential standards" (and, therefore, the operational forms relating to the financial system).[16]

From another perspective, it must be observed that the architecture of the shadow banking system is a prerequisite for new types of risks, even with the effects of a systemic nature. There is no doubt, in fact, that the high degree of private autonomy given to the arrangers allows the application of the latest techniques developed by financial engineering, in terms of process and product.

It is not about proceeding uniquely to the identification of the risks referable to the subjects or activities carried out by the latter, nor examining them as a result of the support activities provided by banks, financial firms, and insurance companies. It is necessary, in fact, to evaluate whether the cases in question concentrate critical factors in "remaining areas," such as to be measured through the evaluation of the so-called "tail risk" (in the same way as extreme events that are not adequately measured by the VaR). Of course, the purpose of adopting specific and effective regulatory safeguards is identified.[17]

It seems that the regulation will reduce the uncertainty, complexity, and systemic risk of the shadow banking system, but it shall not ban—or reserve to banks—the intermediation activity at the basis of the phenomenon. Significant in the field is the position of global regulators, who seem able to promote the recovery—in a context of full legality and oversight—of the shadow credit intermediation process and other activities connected with it or instrumental to it.

This suggests that, at a time of economic difficulty (and expansionary monetary policies), one finds affirmation of the will to preserve the presence of an alternative channel (to the banking one), which increases the cash flows of the real economy and, simultaneously, allows diversification of the portfolios of investors.

Hence, the conclusion is reached that the outcome of a legal analysis of the shadow banking risks shall be manifested in the need for new safeguards that are able to avoid, manage, or mitigate the financial impact of negative events (expected and unexpected). That is, with respect to the provision both of barriers that restrict the entry of risky assets in the equity of the banks and of appropriate infrastructures in support of the distribution of information relating to the shadow banking operations.

It will be up to the sector's authorities to proceed. Firstly, to increase the transparency levels of the phenomenon under observation (together with the correction of any market failures). Secondly, to implement a monitoring system able to apply the results of the studies of "law and economics" to these types of market-based risks. In fact, if the shadow banking system produces effects worthy of protection (by the systems of the democratic countries with a financialized economy), then the regulator authority must necessarily intervene. Such intervention is needed to correct market failures and guide the circulation of wealth towards the well-known purposes of social utility.

In doing so, it will be necessary to bear in mind that, in such a context, the conscious assumption of the risks is at the base of the exercise of free economic activities required by modern democracies. Their actions must responsibly address the execution of transactions to finance the real economy at cheaper prices than those charged by banking intermediaries.

This is one of the reasons that allows consideration of the preservation of this shadow banking. It is linked to the need to activate the safeguards needed to guide its intermediation process toward higher levels of competitiveness, transparency, and stability.[18] Otherwise, it would lead to the denial of a mechanism able to increase liquidity; eventually promoting a "credit crunch" hardly compatible with the current phase of the economic cycle.

Therefore, the reasons underlying the political decision to proceed to a strengthening of supervision on the risks of the shadow banking system appear agreeable. Indeed, once the boundaries of the system are defined and its key vulnerabilities identified, there are no justified reasons—of fairness or efficiency—because the phenomenon should remain completely withdrawn from public supervision, and without the support of the monetary authorities.

This view seems to be supported by the consideration that the equality of citizens (guaranteed by the democratic systems) does not allow a part of wealth to be confined within areas of absolute freedom, even if not illegal (in the presence of the danger that, consequently, crime networks are encouraged). In addition to this are the suggestions of the economic analysis, which orients toward different methodologies able to overcome information asymmetries, conflicts of interest, and other negative conditions that hinder the sustainability of the development of the financial market.[19]

7.4 Operational freedoms inside the shadow banking system

This brings me back to the main question of this chapter: are risks and freedom (to assume risks) two characteristics of the shadow banking system?

In this regard, it may be useful to determine which model should be applied: "supervision per subjects" (used in the ESFS) or "supervision perfunctions" (able to control the activities of market based financing regardless of the legal nature of the subjects involved in their scope).[20] It can also be said that both forms of supervision are able to address and correct the informative asymmetries (and other "market failures"), which hamper the attainment of the *equilibrium* in the shadow banking system. It seems useful, therefore, to verify the suitability of the potential solutions to the organizational risks (of the shadow banking operations) and to those of governance (of the shadow banking entities). This obviously requires a dual inquiry, since both—while interacting with each other—are otherwise influenced by the lack of transparency (the first) and conflicts of interest (seconds).[21]

However, with regard for the risks that originate outside the system, it is necessary to dwell on the interconnections of the shadow banking system with the economic realities financed by it. It will be imperative, therefore, to look at relations with the supervised intermediaries in order to avoid the risks assumed by them being transferred to the shadow banking system (also with simple securitizations).

Similarly, the recession and the process of "premature de-industrialization" can adversely affect the reliability of the underlying securities and, therefore, the functioning of the shadow credit intermediation process. It should also be borne in mind that the non-conventional transactions of central banks—and the interest rates associated with these—can promote levels of competition that can undermine certain forms of market-based financing (in comparison to the supply of credit of traditional banks).

To summarize: risk factors interact in ways that would suggest systemic consequences that must necessarily be subject to a control system in order to enhance financial market resiliency.[22]

7.5 The risks of organizations

From a legal perspective, we will now focus on the area of impact of the risks associated with transformations that are carried out through the shadow credit intermediation process. Indeed, I refer to the business relations between shadow banking entities on one side and banks, financial firms, and insurance companies on the other. An external importance to the risks of organizing the shadow banking operations can be recognized,[23] because the default of the latter corresponds to the non-fulfillment of the obligations contracted (when granting credit lines, insurances, or other benefits) to pursue the modifications of the maturities (i.e. maturity transformation), of liquidity (i.e. liquidity transformation), and of quality (i.e. credit transformation).

Therefore, it is necessary to dwell on the type of risks that qualify the transactions in question, which—as have been seen—are designed to use "short-term deposit-like securities" in order to finance "longer-term loans." In particular, consider the sequence of operations in which the assets are less liquid than the related liabilities. This is a reality that places specific importance on the production process of CDOs, where the form of warehousing is provided in order to add up—in the assets of each vehicle—loans, before, and instruments, after (also referable to a large and diverse set of counterparties).

Consequently, in this case risks exist that are generated during the preliminary stage (before the organizing of the shadow banking operations) when the arrangers start to monitor the CDO demand and select the assets to be placed at the start of the shadow credit intermediation process. However, only after the beginning of the operations does it appear possible to proceed with the measurements in question (not only with regard to maturity transformation, but also the involvement of a plurality of subjects in the same process).[24]

These are, then, issues that do not fit within a business unit, but only in an open market, in which each phase of the process is assigned to a separate legal entity (and the intermediate products are, usually, traded in the financial market).

This is the first characteristic of the phenomenon that must be taken into account in the construction of a system of public supervision, characterized by the need to mitigate the general risk that the financial project is not sustainable (i.e. that it presents minimum probabilities of being able to repay the initial investments).[25] Given this organizing context, there is the need to communicate to the market the trading conditions and, therefore, to ensure full disclosure. This must provide the means of information that can make it possible to know the shadow credit intermediation process and, therefore, to identify the dynamics associated to it.

To the adequacy of the information corresponds, in fact, the reducing of the probability that investors will subscribe to the CDOs without the correct perception of the risk profile that qualifies their content. (Thus, without being in a position to adequately consider the economic and financial value.)

From another perspective, it is necessary to consider—at the beginning of the organization of the shadow credit intermediation process— the variability of the conditions of supply of loans and demand for CDOs. For this reason, the system of supervision should include devices that measure the effects of variability. If possible, it should also introduce elements of correction directly (through the modification of the production process) or indirectly (through the obligation to ensure an alternative operation in case of adverse events).

Complementary to this variability is the need—for the regulator— to avoid the risk that the project reaches levels incompatible with the issuance of financial instruments (liquid and safe) corresponding to the market demand and, therefore, that risks lead to a crisis of the shadow banking system shutting down its brokerage circuit.

Furthermore, the construction of a "deposit-like funding structure" implies the choice of proceeding with the issuance of short-term financial instruments. This exposes the shadow banking system to the danger that a sudden and simultaneous lack (of interest) of investors will be resolved in a disinvestment due—in economic terms—to the phenomenon of "runs" (against which will then be provided specific *backstops*).[26]

This is a particular risk associated with the probability that the counterparties, in the occurrence of certain events, decide not to pursue the investment relationship set in place. Therefore, they seek reimbursement of the amount invested (if possible) or offer for sale (on the market) the financial instruments in their possession. In particular, in the first case, there will be the need to ensure the (safeguard of) restitution. In the second case, it is important to avoid an increase in the supply leading to an unjustified reduction of the market value of the instruments in question.

It goes without saying that the regulator should include other specific safeguards for this matter, including the possible involvement of lenders of last resort. In turn, there is the need for effective devices that are able to mitigate the risks of default of the issuer (if he or she will not be able to liquidate assets in order to repay what required) and of reduction in the fair market value of instruments intermediate to the shadow credit intermediation process. (This would result in damage to the subsequent stages of production of the CDOs.)

Only the first risk can be managed on a bilateral basis (with clauses that prevent an early disinvestment). The second can be mitigated within the market, given the limited effectiveness of authoritative interventions of "suspension of sales" or "disapplication of the accounting criteria" based on the fair value in the financial statements. Obviously, in conditions of complete information, such risks are known to the industry and, at times, are discounted from the prices for the financial instruments issued. In addition to that, another type of risk exists, related to the dangerousness of the forms of organization that can carry out the build-up of high, hidden leverage. Indeed, it seems necessary to introduce regulatory limits to ensure that the entire shadow process shall consist of loan contracts and credit lines (granted to SPVs by banks) not secured by collateral other than the assets used to power the production process of CDOs.

Ultimately, it can be said that the future system of supervision shall, in line with the guidelines of the European Commission, bear in mind that "shadow banking activities can be highly leveraged with collateral funding being churned several times, without being subject to the limits imposed by regulation and supervision."[27]

7.6 The risks of governance

Up to this point, the discussion has focused on the characteristics of shadow operations and the related need for regulation. I will return to that in Chapter 8, in order to consider how the legal system deals (or, rather, is going to deal) with certain operations that can have negative externalities. Before I come to that, this section must consider the organization of shadow banking entities and the risk arising from the choices at the basis of their organizational structures.

Obviously, I refer to the entities that are able to increase the overall risk profile of the shadow credit intermediation process. Many are, in fact, the negative events related to designing errors. These can occur in different ways depending on the type and the temporality in which the activities are produced for credit transformation by certain subjects. It appears, therefore, necessary to dwell on the risks associated with the governance of the entities involved in the initial phase (of granting loans to the real economy), as well as those who realize the securitization and re-securitization (of loans). To these should be added the operators involved in the intermediate stages (with accessory activities, related to the maturity transformation, the credit enhancement, and the leveraging).

Hence, the effectiveness of regulatory solutions that attempt to mitigate the mentioned risks without eliminating the features that allow shadow banking entities to compete with the supervised institutions shall be evaluated. Of central importance are, therefore, the management bodies and internal control systems of the aforementioned subjects. That is, in view of the fact that, in the system under consideration, only the private autonomy can avoid taking excessive exposures (to individual transactions), or the establishment of inappropriate business relations (speculative).

Consequently, it must be taken into account that the risk of governance is determined as a result of a dysfunction of internal controls. In particular, there is the danger that the abovementioned checks are not able to guide management in the request for appropriate services (to banks, financial firms, and insurance companies) and the efficient use of resources. This is a dysfunction that can prevent proper balance (in terms of profitability and sustainability) of the production cycle of the CDOs. Hence, the possibility that a lack of efficiency and guarantee will result in damage, first, of the arrangers and, second, of the whole system.

Moving towards the examination of the risks related to the governance of the SPVs, it should be noted that there is currently a "legal entity typically used to hold securitized assets," to which "a physical manifestation, an office, or employees" does not correspond.[28] Therefore, the establishment of such an entity results in a formal fulfillment, necessary to achieve the benefits of a full autonomy and, therefore, to limit liability in case of default of the transaction. This occurs along with a sharp distinction between the risks of the latter from those of the arrangers, where the creditors of the first can claim the assets of the vehicle, while those of the second can make a claim only on the equity shares.

Moreover, consider that the arrangers may not cover the risks that could remain confined within the liabilities of the vehicle. The fair management of this risk is not due to a legal obligation (related to participation to the equity or the conclusion of contracts), but to a consideration of opportunity (based on the reputational value to be protected by the default of any shadow banking operations). It is important to regulate any spill-over effect, because if the arranger is a bank, then the default of the operation has an impact on an entity able to receive the support of the monetary authorities, by shifting the burden of insolvency from investors to the system.

To conclude on this point, it can be said that (i) the risks of the SPVs do not always produce effects that stay within the relative assets (and, therefore, the legal relationships that are owned by these vehicles); nor (ii) are the errors in the design of the governance the sole cause of transmission of the negative consequences to the supervised credit institutions. There are, in fact, forms of "implicit support of SPVs" due to "potential financial pressures on the sponsor" on the part of the stakeholders who, for various reasons, relate to the latter.[29] And, in the absence of explicit legal relations, it is difficult to foresee the engagement levels of the arrangers or of the sponsors, and thus to quantify the prudential safeguards able to ensure the recovery of the shadow credit intermediation process.

7.7 The impact of European regulation on the shadow banking risks

There are many events that will challenge the capacity of the shadow banking system to absorb shocks arising from financial and economic stress (i.e. of the worsening of market conditions or the difficulty to collect the reimbursement of the underlying loans). It is apparent that these shocks will put to the test the interconnections of the shadow banking system and the banking industry, given the pro-cyclical amplification that occurs at each stage of the shadow credit intermediation process.

However, there is no doubt that, even if the goal is to find a way of reducing accidents (or, rather, managing risks), the forthcoming regulations prohibit the shadow banking entities from taking part in the activities linked to banking and regulated financial markets.[30] And the same is true for the banks, financial intermediaries, and insurance companies that—according to the rules laid down by their relevant regulatory framework—provide services to these entities. Indeed, it should be taken into account that a process of evaluation is in addition to any exposure of these supervised companies (to the risks of the shadow credit intermediation process), and that this evaluation leads to the provision of adequate amounts within the technical reserves (according to prudential criteria that ensure the stability and solvency in time of the same).

This means that the management (and the business function in charge of the internal controls) of any supervised institution is entrusted with the task of avoiding excessive exposure (towards individual shadow credit intermediation process) or improper business relationships (with SPVs deprived of adequate capital support). Therefore, the focus should fall on the prudential regulation of the securitization transactions brought by EU Regulation no. 575/2013 (so called CRR), which leads, first, to the assessment of the retention of net economic interest (of the loan originator) as a safeguard towards the shifting of risks from the banking sector to the shadow.[31] To this corresponds the ability to accurately determine the coverage of risks related to securitization transactions involving the use of SPVs, to which must necessarily be entrusted the task of defining the integrated process of management of all the issues: current and future.

In the end, the entire shadow credit intermediation process appears to be exposed—beyond the credit risk that qualifies the loans issued to the real economy—also to other hazards related to the designing and management of operations. A risk of governance has been found, which is not imposed only on entities (or, rather, on its direct shareholders and stakeholders), but also on all those involved in the process. Hence, the need for appropriate forms of external control, apart from a system of information storage, accessible to all stakeholders. It follows that the use of risk mitigation instruments can facilitate the work of the arrangers, but it should not be the only element capable of ensuring the quality of the operations.

7.8 The risks of the entities "too big to fail"

The situation seems quite different when dealing with companies that are "too big to fail." With particular regard for their counterparty risk, it must be pointed out that the design of the operation may provide a contractual netting agreement, which in the intermediate stages of the shadow credit intermediation process may refer the mitigation of risks to the extinction of mutual exposures between entities involved in several connected transactions.

The possibility must be considered, then, of further problems in the event of intervention by financial firms classed as "too big": defined as such when their distress or disorderly failure would significantly disrupt the wider financial system and economic activity.[32] Indeed, the services offered by these companies are targeted at a wide audience of recipients, such as to indicate further their characteristic of being "too interconnected to fail."

This helps in sharing the concern of the FSB for the effects of operations that are outwith prudential standards. Achieving more pervasive forms of supervision, in order to determine more caution in banking (and at the same time, a supply of equity), can avoid eventual crisis in the whole financial system. In this case, the possibility of preventing such G-SIBs from participating in the shadow banking system should not be excluded, because of the existing asymmetry between the SIFIs and other entities with whom they come in contact (in terms of operational capacity, market power, and bargaining power).

Therefore, it may be concluded that the supervision must take into account the risks—of the shadow banking system—related both to the design of the production process of CDOs and the governance structure of operators who participate. And, in doing so, it must take into account the fact that the size of the entities (or the one of the counterpart) can amplify the problems resulting from an abnormal development of the loans.

Hence, we cannot underestimate the direct consequences of moral hazard. This applies, in particular, to the risk that the arrangers realize excessive sophistication of the operations or of the process (which does not resolve in an effective credit enhancement, but in an appearance in accounting intended to provide evidence of its inconsistency in the long term). This is a risk that arises when those who realize this stage do not suffer the negative consequences of the events of default (which can only happen after a certain period of time).

In the absence of the traditional safeguards designed to limit moral hazard, the phenomenon in question has criticalities that the market alone cannot solve. And this is so not only for the conditions of opacity that qualify the case under observation, but also for the possibility that certain operators accept excessive levels of risk taking. In this case, undoubtedly, the dangers are not in line with the economic incentives that motivate the behavior of the industry. This shows a context in which, even today, the operational freedom allows the possibility of failures incompatible with the conditions of stability and security needed by the capitalist economy.

Despite this, it is easy to feel the need for public intervention that would submit to specific controls the shadow banking entities, preventing the temporary nature of their participation in the production process. Thus, an effective regulatory framework becomes necessary, enabling them to withstand the negative effects of risks to the people who have started them, even allowing solutions of private law, which give the possibility of action (for damages) in order to attack the personal assets of the latter.

7.9 The exogenous risks

The traditional approach ends the analysis of shadow banking risks, before considering the global financial environment hosting the shadow credit intermediation process. Although this is understandable, it seems to be too simplistic. The idea *a priori* that the shadow banking system is also exposed to the negative effects of events external to its production process and, in particular, to the effects of the performance of the (regular, but alternative to it) financial and banking systems cannot be rejected. And this is the case both in terms of price and quantity.

Moreover, we shall consider that the system's architecture is built according to disciplinary models that allow the immediate passage "from credit losses to counterparty contagion to crisis." This is on the grounds that "the excesses of exuberant risky loan origination, esoteric off-balance-sheet entities... and other complex instruments and entities involved in securitization embody the Minsky progression of financial market risk taking," witnessed by "a succession of events leading to dysfunctional capital and money markets [that] began as early as summer 2007." Hence, the interpretative guidance that "bad loan underwriting contributed to the overabundance of cheap credit through 2007." This corresponds to the conclusion that, "there is a distinctly human element to the systemic build-up and subsequent systemic run on the shadow banking system."[33]

Therefore, each external event able to reduce the amount of capital flows invested in the shadow banking system is resolved in a risk for the system under observation. This suggests that the shadow—as in traditional financial markets—is vulnerable to fraud, human error, disruption, errors in exchange systems, and breach of contract.

Furthermore, it should be considered that capital flows can also be reduced as a result of events depending on market trends and, in particular, on the conditions of supply of financing or safe and liquid instruments. In other words, the phenomenon in question is exposed to the risk that the traditional system attracts all available resources, thereby preventing the smooth running of the process.[34]

In this regard, note the possibility for banks to offer loans on favorable terms. Upstream of the latter there may be monetary policies that, as well as providing a lowering of interest rates, grant collateral loans to businesses.[35] This is a situation of competitive advantage for traditional credit institutions in comparison to the entities operating in the shadows, which do not have direct access to the funding of the monetary authorities.[36]

As will be seen in Chapter 8, the result of such a policy is that the flows at the beginning of the shadow credit intermediation process shall reduce. This is because of the interest rates charged by banks (which will erode the competitive strength of the shares of market for loan originators) and the chance given to intermediaries (which operate on the model Originated to distribute) to refinance themselves with the monetary authorities under more favorable terms than those offered by the market.

In light of the foregoing, there seems to be justification for a public intervention aimed at regulating the manner in which a shadow banking operation can use resources from traditional banks, and then transfer its risk of insolvency to the latter. This intervention shall manage the risk arising from the arrangers that use the financial leverage, not only in order to increase the volumes of its activity, but also to secure a source of liquidity in case of runs (and, therefore, to be able to cope with the "reimbursement requests" through the activation of traditional credit lines).

Undoubtably, this is a situation in which the leverage, on the one hand, increases the ability of the shadow banking system to finance the real economy and, on the other hand, amplifies the effects of the risks under examination. It will, in fact, be necessary to avoid placing in the hands of the bank the negative effects of poorly-designed shadow banking operations. However, the regulator shall take into account that risks associated with the use of leveraging can make use of those quantitative remedies by introducing a "ceiling on loans" (intended to be an administrative tool to control credit) or limiting the volumes of activities to a multiple of the capital base of the vehicle (according to the model laid at the basis of the Basel Accords).[37] It should be noted, however, that these solutions have not so far performed well outside the traditional banking market.[38]

Obviously, the choice to intervene through the application of the containment mechanisms of leverage may involve a high transaction cost, due to the market-based nature of the phenomenon and the large number of subjects involved in each shadow banking operation. Despite this, such an intervention shall avoid any uncritical extension of the existing banking prudential rules, because it should reduce the aforementioned competitive drive of the shadows.

Conversely, the decision to increase the levels of transparency (of the shadow banking operations) and the construction of a mechanism for centralization and dissemination of information (necessary to establish equilibrate investment choices) can record the risks of the system under examination. In this case, consideration must be given to the conservation of the features of alternative nature and competitiveness that characterize the essence of shadow banking (or, rather, the complementary function to the banking sector in the circulation of wealth and financing of the real economy).

In addition, with adequate levels of disclosure, supervisory authorities will assess whether the supervised entities involved in shadow banking operations have properly weighed-up the risks assumed. If so, transactions with the shadow banking entities will have been prevented from translating the negative effects of poor planning on the part of banks, financial firms, and insurance companies.

With particular regard for the securitization transactions, it is necessary to bear in mind the effects of warehousing. There is no doubt that the variety of obligations underlying an issue of ABCP and ABS should be positively evaluated, if considered with regard to diversification and, therefore, to the fact that a single event of default has little impact compared to the value of a single operation.[39] It cannot, however, be forgotten that, in this case, the diversification increases the likelihood of encountering an event of default (given the abundance of the abovementioned obligations) and, therefore, leads to the consideration not so much of the *an* as much as the *quantum* of the negative effect associated to it.

In this context, note the praxis to manage these risks through a division into classes of financial instruments. Hence, the possibility that—according to the regulation of the issuance—there is a right to a "priority of claim" for some of them. Consequently, the risk of loss is reduced for participants who have recognized this right, and at the same time increases for those who sign the instruments of a lower class. Not always, however, is this form of risk segregation able to ensure fully the reliability of the entire shadow credit intermediation process. Indeed, the activity of warehousing leads to a mixture of assets that results in a thorough understanding of what underlies each issue (and, more importantly, prevents a proper evaluation of the instruments issued).

Therefore, it is possible to identify the risk of a mispricing and, in turn, the danger of an inefficient allocation of the risks and available resources.

Again, the research into the best ways to ensure the safety of the shadow banking system and the preservation of its characteristics (diversity from the banking sector) suggests the need for public intervention that will increase the level of transparency and improve the quality of financial reporting. Hence, there is the need to regulate the communication activities of the shadow banking entities and to require supervised subjects to gain the knowledge necessary for a conscious investment of their capital.[40]

From another perspective, it should be noted that specific risks apply to the derivative transactions. If the use of futures, options, and interest rate swaps allows completion of the design—of the shadow credit intermediation process—with elements of detail that can limit the effects of the variability of certain components, then, it introduces an additional risk associated with the reliability of the subject who promises the operation raised in the derivative contract.

At the same time, in the design of a shadow banking operation, the decision to negotiate a derivative in a regulated market (or outside of it and in the absence of a central counterparty guarantee) may give rise to additional risks, other than those which are the subject of the contract. Consequently, I feel the need for a rule that is able to prevent derivatives replacing the aforementioned risks (linked to the counterparty creditworthiness) and thus preventing the same from being resolved in a mere risk transformation, rather than a real mitigation activity.

Another risk profile concerns the possible elusive nature of the shadow banking operations and, therefore, the possibility that they are incorporated in a given country for the sole purpose of obtaining the benefits of a circumvention of rules and, then, a regulatory arbitrage. Such conduct is not considered legitimate by democratic systems, so that the operation will have no legal effect, or at least will be ineffective with respect to third parties.[41]

To conclude on this point, it should be emphasized that—in the operations in question—the most obvious risk is that investors do not have the data available and the information necessary to make estimates and forecasts on the quality of the CDOs, and predictions on the trend of the loans underlying them. Hence, the impossibility of a conscious acceptance of a certain level of danger of the shadow banking operations.

That is, in the absence of safeguards designed to avoid the pursuit of profit leading the parties to operate in conditions of uncertainty.

It goes without saying that the opacity of the shadow banking system makes it difficult to complete a full and preventive mapping of its financial risks. Moreover, the sequential nature of the shadow processes implies, in itself, the absence of a subject that knows all the activities and relationships that give content to the shadow credit intermediation process. This suggests the importance of a support system capable of ensuring the centralization of information, to be supplied with relevant data and news that—after being collected through communication, by the shadow banking entities, of specific statistical reports to the competent authorities—should be established through an effective monitoring action.

This leads to the hope that regulatory intervention promoted by the supervising authorities is not limited to extending the traditional provisions established for banks, or to establishing a prudential control designed to assess in conventional ways the adequacy of the shadow banking entities. Indeed, it seems necessary to follow the guidelines of global regulators, which promote rules of risk management based on the transparency of information and on ethical behavior, in order to ensure the sustainability of the phenomenon under observation.

7.10 The particular implication of monetary policies

It is important to realize that both the cost and the benefits of the competition between the alternative systems of banking are linked to the monetary phenomena. It would be more important to understand that the actions taken by central banks—and, in particular, those which aim to achieve an expansion of the monetary base by increasing resources for provision of loans (which, usually, associates also with a reduction in interest rates)—can influence the market trends and, in turn, the interest of the loans (to be securitized) and the yield of the CDOs.

Hence, the discussion of the risks exogenous to the shadow banking system shall include the impact of the expansionary monetary policies on the demand and supply that goes to its market, because of the risk that such a policy will reduce the volumes of the system in question.

This brings to attention the Decision ECB/2014/34, relating to "transactions aimed at refinancing the longer term" (so-called TLTROs). These operations were designed to improve the functioning of the transmission mechanism of monetary policy by supporting the credit to the real economy.

It goes without saying that, in this case, the objective pursued by the ECB has been that of pursuing its price stability mandate (art. 127.2 of the Treaty on the Functioning of the European Union and art. 18.1 of the Statute of the European System of Central Banks and of the European Central Bank), with specific reference to the contrast of the financial crisis. But, the preferred type of collateral allows you to achieve more goals. This applies, in particular, to the stated intent of the Governing Council "to support bank lending to the non-financial private sector, meaning households and non-financial corporations, in Member States whose currency is the euro."[42]

More generally, a diversity of approaches of the monetary authorities (ECB and Fed) can be observed in the definition of the rules governing the eligibility of collateral, demonstrated by a recent study by the Bank for International Settlements. That is, in terms not only of eligible asset types, but also of other dimensions such as eligibility across lending facilities, haircut policies, collateral management (earmarked or in a pool), and so on. Indeed, the analysis of the collateral rules and practices applying to standard open market operations (OMOs) and standing liquidity facilities (SFs) of each central bank shows divergent lines of intervention: uniform versus differentiated (operations/facilities); narrow versus wide (asset classes); earmarked versus pooled (loans).[43]

Specifically, the approach shows particular interest in the Bank of England's approach, which became differentiated only since June 2007 where—in terms of issuer type—it varies with facility, and the collateral delivered is earmarked for specific loans or repos. Similarly, that can also be said for the FED (herein referred to as differentiated): narrow for OMOs and wide for SF, earmarked for OMOs and pooled for SF.[44]

At the quantitative level, the empirical evidence seems to show that the presence of loans is a minority in the composition of collateral pledged to central banks. This leads to the decrease of liquidity (i.e. financial instruments offered as collateral to central banks) during periods of tension of markets. Thus, there is in this case a new type of risk arising from the fact that one longer-term influence comes from the design of central bank collateral policies, since collateral use at central banks influences collateral practices in the market.[45]

Moreover, this risk seems able to also influence the policies of the Bank of England, which takes as guarantee portfolios of certain types of loans to non-banks in its Discount Window Facility, given—according to the aforesaid analysis—its (main) purpose to support financial stability by acting as a backstop provider of liquidity insurance to the UK banking system. This proposition affects the size of the "bank's list of eligible collateral." Not surprisingly, in an authoritative technical context, it was about to be emphasized that this list, "which has already been expanded significantly in recent years, will be extended further to include the drawn portions of corporate revolving credit facilities."[46]

In other words, it can be said that one of the most threatening risks (external to the operation of the shadow banking system) is the effect of the (expansionary) monetary policies. These policies not only enhance the operational capabilities of the banks, but they are able to guide the activities towards the granting of credit to the real economy at ridiculously low rates.[47] (High) quantity and (low) prices, then, are the determinants of a risk that is manifested by the decrease in flows that feed the shadow process with secure loans (since—as has just been shown—these can be used as collateral to access financing from the central banks).

It seems, therefore, possible that the monetary authorities can use their tools to contain the expansion of the shadow banking system. This is, obviously, an authoritative intervention that implies the endogenous realignment of the relevant incentives, such as to orient the demand for loans towards credit institutions that fall within the scope of the public supervision (and benefit from the related guarantees).

8
The Shadow Banking System and the Need for Supervision

In this chapter I will move from the link between the deregulation of the banking industry and the freedom to perform market-based financing, in order to understand the need for checks and balances in the shadow banking system.

In this context, I take into account the effects of the recent evolution of the EU supervision and its limits in controlling certain financial operations. This is why attention will be focused on the effects of the recent developments in the EU financial market, caused by the implementation of both the ESFS and the EBU (with its single supervision, single resolution, and single rulebook).

8.1 Checks and balances in the shadow banking system

The current set up of the banking industry and financial markets—beyond the redefinition of the powers of national supervisors and of the roles of the monetary authorities (in particular, of that assumed by the ECB after the establishment of the EBU)—is based on a substantial *freedom* of the financial intermediaries and the companies that manage the trading venues.[1] This is the result of the old deregulation process (of capital markets), when the orientation towards the "neutrality of public intervention" began to correspond to a revision of the special rules governing the movement of capitals and, in the European context, the harmonization of a legal framework intended to implement a single market.[2] Undoubtedly, the need to promote the advancement of productivity led national regulators to build, in private form, the propulsive acts of the exchanging activity. Previously, however, the duty to control the financial markets had been deferred to public authorities.³ It should indeed be noted that the situation prior to Directives nos. 6 and 93 was evaluated critically, also with regard to the monopolistic configuration of regulated markets.⁴ To this analysis corresponded, in fact, doubts about the coherence of the European system of markets, increasingly questioned due to the implementation of the principle of freedom of capital circulation and, in particular, the start of the globalization of finance.⁵

At present, then, the combination of technical ability and professionalism ensures the orderly development of trading and, therefore, the optimization of the relationship between the efficient allocation of money, the effective development of economic and social welfare, and the equitable distribution of the wealth produced.⁶ In this context, the *creative flair* of certain financial engineers designed the key elements of the shadow banking system.⁷

We shall consider that at the basis of the current configuration of the financial markets there is a system of self-management that—from its early stages of application—has played a complementary role to that of supervisors, but not a role of "complete replacement."⁸ The possibility of a government intervention in defense of collective interests, such as to protect the transparency of information and, in general, to correct any market failures, has not been refuted. In other words, financial liberalization (and deregulation of exchanges) should be credited for having allowed the extension of the operational capabilities of intermediaries, the improvement of the production function of the markets, and, consequently, a more efficient allocation of available resources for the financing of the real economy.

In emphasizing this point of view, I shall now observe an important point of agreement: to the configuration of the market as a private enterprise corresponds a new business activity of organizing venues and managing exchanges. In addition, the "competitive regime" corresponds to the proliferation of centers and, therefore, the segmentation of the operations in more the one trading venue. Finally, to the overcoming of the "principle of concentration of trading on the stock exchange" (Directive 2004/39/EC) corresponds the fragmentation of information on quantities and prices of the securities in circulation.⁹

Furthermore, it is apparent that this configuration has made possible the development of new operating modalities (from the securitization of loans to derivative finance) and the application of the techniques of "high frequency trading" and "algorithmic trading" (which complements the automatic formulation of orders and the contemporary entering of these in multiple locations, only to conclude the one order that finds the more convenient response).[10] In this context it should be noted, also, that the recognition of "wide open spaces" to private autonomy and self-regulation has made it possible—for the financial operators—to replicate, through a series of market transactions, the banking activities traditionally carried out by banks.[11] Thus, the efficiency gains (due to the liberalization) have allowed the markets to compete with the banks in terms of rates and volumes.[12]

The result is a system in which trade liberalization not only allows the transfer of wealth from operators in surplus to those in deficit, but also makes it possible to realize effects of leverage, maturity transformation, and credit enhancement. And, as has been seen, these effects can be realized in the market; hence, the logical and operational premise for the success of the shadow banking system.

As an initial conclusion, it must be highlighted that the market-based intermediation takes place in a context that, on the one hand, is outside the scope of banking regulation supervision and, on the other hand, is subject to a legal system that safeguards the smooth functioning of trading and the protection of investors. That said, it should be noted that the recent financial crisis has highlighted the limits of such a control scheme, given the negative consequences that have been recorded since the subprime mortgage crisis. As has been said many times already in the course of this book, it is now a common and accepted idea that the shadow banking system should be subject to monitoring and, in turn, a more pervasive form of supervision, able to drive the evolution of the phenomenon to the right levels of stability and security.[13]

8.2 Economic determinants of the supervisory system on shadow banking

I can best introduce what has to be said by considering that the "short twentieth century" ended with an unjustified asymmetry between (i) government supervision in banking (and on regulated trading venues) and (ii) the absence of forms of control on market-based mechanisms of credit intermediation.[14] Therefore, the state protected only the depositors (who turned to banks) and, in less intense modalities, the investors who accessed directly regulated markets of financial instruments.[15]

These were the effects of a "political choice," which made the banking sector safer than others (even if the "bail-in" of Directive 2014/59/EU will overcome this choice) and a high level of protection of savings was linked to the further aim of controlling the operations of credit.¹⁶ This choice also matched a stable equilibrium and, in particular, guaranteed the smooth functioning of the credit circuits.¹⁷

Hence, after the deregulation of the 1990s and before the crisis of the 2000s, the success of the shadow banking system was due to spontaneous forces (of the market) and free investment decisions (of investors). Both used specific (market-based intermediation) mechanisms to obtain a positive economic result even in the absence of constraints to private activity and a specific inspection system.¹⁸

Another determinant of the aforementioned success is linked to the financial globalization, given the asymmetry between the scope of the supervisory authorities and the increasing importance of cross-border transactions. New freedoms arise from the limited power of national (and regional) supervisors (closed in a single territory, too little if compared to the vastness of the financial system), together with the possibility to take advantage of the aforementioned asymmetries in an opportunistic way.¹⁹ This determinant influences, in particular, the intermediate stages of the shadow credit intermediation process that can be run in countries without the traditional safeguards for the protection of collective interests. It leads, in this case, to the risk of having a business made of operations in which the (individual) pursuit of profits is exaggerated—social utility is compromised.²⁰

It could be concluded that the current regulatory choices lead to lack of supervision on the shadow banking system. On the contrary, however, there appear to be no socio-economic justifications for this absence of public safeguards. It can be deduced from this first conclusion that there is a need for preventing a lack of organization of the "supervision apparatus" allowing the shadow entities to proceed with the deliberate assumption of high levels of risk, which are incompatible with the overall stability of the financial system. And this is the result of both economic efficiency and social fairness.²¹

Following the current trends of the capital markets, it is important to highlight the effects of TLTROs on the shadow banking system: given that the ECB can invest also in certain securities issued through any of the shadow banking operations, and that the ECB can improve (or reduce) the performance of CDOs (which can be used as collateral for these refinancing operations).²² According to the analyses of the FED and the Bundesbank, it is easy to understand that the shadow banking system can be crucial for the smooth functioning of the transmission mechanism of this monetary policy. In this context, there is a need for new rules, which are able to prevent the production of "toxic assets" that, actually, do not comply with the procedures of the Asset Quality Review (designed to prevent the supervised intermediaries exposing savings they have collected to excessive risk).[23]

Thus, there is no justification for the absence of regulation and the lack of supervision over the shadow credit intermediation process. The focus here shall, therefore, be on the guidelines provided by the competent authorities to introduce public controls, proportioned with the methodology of intermediation and the type of resources that give content to each stage of the shadow credit intermediation process.

8.3 The shadow banking system in the European internal market

I have mentioned in the previous chapters that the further evolutions of the European Union might be more favorable to the supervision on the shadow banking system than the previous regime has been. Moving on to the analysis of the recent trends of European financial market harmonization and integration, it is worth repeating that—from an operative perspective—a clear separation between the shadows and the regulated does not exist, as the EU regulation (of standards provided to achieve the free circulation of capital) ends up interacting with the whole financial market, but not preventing the relations between supervised companies and shadow entities (see Table 8.1).

Table 8.1 Supervising network

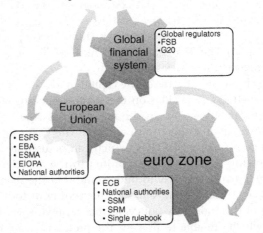

In fact, in many cases, European directives have created supranational rules for the competition in setting up the shadow banking operations carried out in the EU territory, in line with the structure of the European market.[24] Despite this, at the beginning of this millennium, European rules—in providing the notification to the Commission of any authorization to carry out banking activity (and the list of authorized persons, art. 11, Directive 2000/12/EC)—ended with the preference for an open market, in order to avoid barriers to entry that are based on alleged "economic needs of the market" (already overcome by Directive 77/780/EC). If, on the one hand, the credit market was opened up to internal competition (including banks), then, on the other hand, the European regulator delimited the boundaries of the shadow banking system. (That can be seen from the relationship between the "systemically important financial institutions" and complex regulation concerning supervisory actions.)[25] At the same time, the European regulator set a standard that was not overcome by the following restatements. This is why the structure of Directive no. 2013/36/EU and Regulation 575/2013 confirm this regulatory framework (given the implementation of the developments reached in the Basel Accords).[26]

However, doubt cannot be cast on the fact that the European regulator had not been able to eliminate the elements of instability associated with the financial market, leaving without prejudice the conditions that, in the past, justified the intervention of the national financial supervision of banks.[27] This applies, in particular, to the opacity of the process of maturity transformation and the ability to expand business volumes over the extent necessary to fund the credit to the real economy.[28]

Therefore, the proper functioning of the European internal market requires further measures designed to mitigate the negative consequences of certain shadow operations and, hence, a more pervasive control (more than just a system of monitoring). Undoubtedly, in this way, public intervention will be able to ensure the absence of barriers to the entry and, therefore, competitive conditions that secure the effectiveness of the safe equilibrium required by the European Union for its markets.

I shall conclude, one more time, that there are economic determinants for the regulation of the shadow banking system, given that also the financial crisis—in intensifying the opportunities for interaction and dialogue between policy and administration—will constitute a factor of acceleration of the construction of the European single capital market. In particular, an intervention is expected, aimed at stabilizing the certain supervisory devices that, now, ensure the regularity of finance and the effectiveness of its relationship with the real economy.

8.4 The role of European institutions

Worthy of special attention is the analysis of the European institutions' position in promoting an expansion of the areas of public supervision (that would reflect the directions of the G20). We shall consider, then, the potential to include the shadow banking system within the scope of the European supervision.[29]

It should be noted that the new regulatory trends appear able to go beyond the obvious limits of a process of harmonization that did not (and maybe still has not) reach the level of integration necessary to ensure the safety of the "EU internal financial market." The process then faced negative effects (of regulatory arbitrage) on the free circulation of capital and on the security of the modalities in which Europe's wealth is invested. In particular, the need was felt by many to complete the regulatory action with regard to the necessity of introducing new control authorities over the European market (surpassing the previous, incomplete systems based on evanescent forms of operational links between national authorities). And this seems to avoid the fact that a lack of functionality of the controls would be resolved in a reduction of the strength of the actions taken so far to address the negative effects of the turmoil in question practiced in banking, finance, and insurance.[30]

We are dealing with regulatory choices that shall be in line with the recommendations of the De Larosière Report and the following criteria contained in EU Regulations nos. 1022 and 1024 of 2013. These choices innovate European policy of prudential supervision on credit institutions, creating the possibility to entrust the ECB with specific tasks, wide enough to take action on the existing interrelationship between the shadow banking and the regulated market.

It goes without saying that, in the outlined context, the existing asymmetry must be taken into account—in regulation—between the EU and EMU areas. This is because the internal capital market benefits from the most innovative actions, but only within the euro zone (and the member states that have established a "close cooperation" with the euro-system).[31]

Therefore, specific importance is assumed by the subscription to the SCG Treaty and the ESM Treaty; both only relevant to member states that have adopted the euro. These countries, as well as participating in the directive bodies of the ECB, are gathered in new bodies—including the so-called "Euro Summit"—able to promote new integrated forms of currency and economy government, as well as the rescue of states in difficulty.[32]

Despite this, it isn't yet clear if the new legal order of the EU internal market will be able to ensure a smooth adjustment of finance and, therefore, follow the process of economic integration of the continent targeted by the legislator.

Hence, I shall emphasize the uncertainty concerning the possibility of parallel structures (related to trading that takes in place in the European Union and that attributable to the euro zone). This also implies uncertainty of the boundaries of the shadow banking system, as the opacity cannot be reduced (rather, amplified) that—as we have seen—characterizes the interconnections between the latter and the traditional banking system.

8.5 New supervision on the shadow banking system in the European Union: the action of the European Commission

One of the further and deeper issues, however, is the way that the European Commission followed to limit the risks that arise within the shadow banking system. In particular, one must clarify if the recent rules—adopted (since the early stages of the financial crisis) to ensure the accuracy of securitization and, more generally, the implementation of the political option to increase government intervention in the European market—are useful to create transparency in the whole financial market.

The assessment of the lack of consistency in the supervisory practices (among different jurisdictions) exceeded the Commission's original goal of ensuring competition in the market (under the responsibility of DG Competition), in order to introduce a system of controls to mitigate "the risks related to collateral reinvestment and borrower/lender default in securities lending."[33] Hence, we can understand the meaning of the provision of a new DG on "Financial Stability, Financial Services and Capital Markets Union," which will have the task of "ensuring the Commission remains active and vigilant in implementing the new supervisory and resolution rules for banks."[34]

However, the need to correlate the intensification of financial supervision (of responsibility of the Commission) is addressed with the introduction of new principals in order to supervise the shadow banking system and its expansion velocity, determined by the European Economic and Social Committee as an important factor of systemic risk.[35]

It is acceptable that, in the absence of such interventions, the regulatory asymmetries (between the banking system and the shadow) would increase and the question asked by the Commission remains unanswered: "how can we ensure that repo markets are safe and what regulatory and supervisory tools can be used to enforce this?"[36] Therefore, I shall verify if these new forms of supervision will guide the European institutions towards the goal of increasing the disclosure (and, consequently, the transparency of the shadow banking system). Thus, the importance of the *Green Paper on Shadow Banking* and the related proposal for the oversight of the phenomenon can be understood. However, I shall look at this paper as a draft for a regulatory intervention that, on the one hand, corrects market failures and, on the other hand, facilitates the dissemination of the information necessary to ensure a conscious risk taking. It is clear, then, that the purpose of monitoring the circulation of shadow banking risks is to contain its negative effects and, ultimately, avoid their shift towards regulated markets (i.e. their landing in the balance sheets of supervised entities).

At the same time, the action plan of the Commission shows a comprehensive approach to the problems of the shadow banking system, taking into account the trading venues located in industrialized and emerging countries (regardless of their settlement in Europe) given that, in any case, the shadow banking operations should produce effects on the internal market.[37] Hence, there is a belief in the possibility of the European Commission's leadership in the regulation process of the shadow banking system.

These shall be the effects of a new EU Roadmap, able to develop a comprehensive financial reform, in order to achieve the goals outlined in the G20. This does not mean, unfortunately, that the process of defining the legislative process should enable the Council and the European Parliament to adopt a regulatory framework for the system under consideration. There is still the need to identify the obstacles to a full Capital Markets Union.[38] In addition, there appears to be quite substantial progress on the completion of the anti-crisis action, which started with the strengthening of capital requirements (Regulation EU No. 575 of 2013) and the introduction of new rules for derivative OTCs (Regulation EU 648 of 2012), as well as other safeguards to ensure that the financial system, intermediaries, and markets are properly controlled (Directive 2014/65/EU).

That said, it is useful to point out that the Commission's objective shall be to improve—in conjunction with the competition, also—the resilience of the European financial system to future shocks, internal or external.[39] Therefore, the regulation of the shadow banking system appears to be set to influence all the operations that affect Europe, even if the activities are carried out beyond the borders of the TU internal market. This suggests the importance of a high-quality regulatory complex, able to support also the relations with third-party countries, in order to avoid the presence of "legal havens" in which the riskier stages of the shadow credit intermediation process are carried out.

8.6 The action of the European System of Financial Supervision

It may be convenient, for the purposes of this analysis, to go deeper into the interpretation the regulatory effects of the new system of European financial supervision and the analysis of the impact of certain applicative solutions (adopted in the aftermath of the crisis), in order to identify what might be the future balances (between freedom and control) in the shadow banking system.

This assessment shall aim at understanding which results—of the new European System of Financial Supervision (ESFS)—condition the performance of the process and, at the same time, which supervisory practices must still be extended to this sector. It is known that the establishment of the ESFS was introduced in order to reduce the degenerative factors of advanced capitalism, which prevented conciliate solutions between development and equality within the European Union.[40] There is no doubt that this approach was intended to ensure the protection of the economic process and, therefore, the contrast of the recent crisis' effects. Thus, it is not surprising that the new supervisory bodies, after observing the cross-border transactions that gave content to the European financial market, perceived the need to introduce more pervasive rules in an industry that—as verified—appeared to be at risk of misuse (by operators focused on speculative practices).[41]

All in all, EU regulation shall not be concerned only (or even primarily) with the reduction of the costs, but also with "shaping tastes" and the need for an equal distribution of the opportunities.[42]

In this context, the focus is on the action of the European Systemic Risk Board (ESRB), aimed at controlling the interconnections able to transfer the risks from the shadows to the (systemically important) intermediaries. This applies, in particular, to the establishment of a framework of macro-prudential supervision promoted by independent bodies of the European Union in charge of the macro-prudential oversight of the EU financial system (i.e. ESRB and European Supervisory Authorities (ESAs)).[43]

There is currently not just a mere promotion of a "safety net" (responsible for the containment of the interconnections that can undermine the stability and confidence of the whole European financial system), but there shall be a set of guidelines (intended for the supervisory authority) to promote the use of monitoring tools (macro) introduced by EU legislation. At the same time, it is expected that there will be certain analytical criteria useful to identify systemic threats, proceeding, therefore, "beyond banks to include the insurance sector, shadow banking and financial infrastructures."[44]

In addition, the ESRB will be expected to perform the task of monitoring the circulation of wealth, in order to direct available resources towards safe and efficient delivery systems from a macroeconomic point of view. This shall be, however, a short-term objective for those who still today face a reality in transformation, which has not yet identified the mechanisms needed to manage the instability of the shadow credit intermediation process and its aggregate effects.[45]

However, the phase of creating sustainable pathways for economic recovery will not be quick; and the set-up of the oversight will require specific efforts by the authorities responsible for EU supervision. If, as it has been shown in the first analysis on the subject, the structure of the three ESAs should offer a renewed configuration of the institutions in the financial sector (surpassing in breadth and depth of action the previous system),[46] it is evident how the ESRB should, conversely, take into account also the shadow banking system (in order to pursue the goal of a more effective and uniform control of the sector within the European Union). If not, it would leave unsatisfied the need to identify a homogeneous supervision of the financial cross-border issues, such as dealing with the waves of instability that spread through the known network of global interconnections.

Ultimately, this doesn't mean including the shadow banking system in one of the areas of macro-prudential or micro-prudential supervision. Rather, it means entrusting the (network of) national and European authorities responsible for the adoption of the rules necessary to introduce a form of control appropriate to monitor risks and to ensure the correctness of the operations that give content to the shadow process. This creates, therefore, a task that goes beyond the duty of sincere cooperation (between the administrations concerned), mentioned in the European regulations.[47]

8.7 The responsibilities of the European Supervising Authorities

Before turning to the latter evolutionary trends (arising from the EBU and the global regulators), we must deal with the fact that (in the European Union) the centralization of the functions in the hands of the ESAs is not able to guarantee the effectiveness of supervision; and indeed, the involvement of national authorities does not connect perfectly with the objective of ensuring the functioning of the internal market through a uniform surveillance market. In this case, the policy instruments available to the ESAs appear to have limited effect on the shadow banking system, especially when evaluated with regard to their amenability to measures to promote common practices and draft regulatory technical standards (regulatory and of implementation, which can then be adopted by the Commission with the value required according to art. 288 TFEU) or, alternatively, to guidelines and recommendations addressed to competent authorities or financial institutions.

At the operational level, it appears necessary to have a mechanism that allows intervention not only in cases where there is a negligence or an incorrect application of EU law, but also in cases in which the European perspective is necessary to manage the degenerative forms of shadow banking. Accordingly, it can be argued that the task of making a "census" of the entities a strict monitoring of operations should be attributed to ESAs, as well as functions that cannot be confined only at a "soft law" level.

The preference can be highlighted for a European system of financial supervision that takes into account the transactions (between banks and non-banks) on a global basis, since that commitment is justified by the gains in efficiency (control) and safety (in exchanges).[48] Thus, there is the need for close collaboration also with the authorities of third-party countries (in order to build a common database, at least in terms of indexing and accessibility, and mapping of the interconnectedness of the shadow banking system with the rest of the financial sector).[49]

It is evident that high-level supervision is needed. It is from the top, in fact, that one can see the full extent of the risks embedded in "hidden credit intermediation chains," and their systemic importance. Only a "globalized vision" can fulfill the expectations of the G20 and the FSB, given the idea that "regulatory measures should be targeted, proportionate, forward-looking and adaptable, effective, and should be subject to assessment and review."[50]

Therefore, the ESFS shall face the challenge of improving the transparency in the financial markets through actions that must necessarily (i) go beyond the traditional boundaries (of the European territory) and, at the same time, (ii) combine actions that are both indirect (carried out through the imposition of specific obligations to the credit institutions involved in the shadow banking operations) and direct (by extending the scope of the European financial system to the shadow banking entities or by applying new rules to them and to their activities). Both these actions will have an impact on the entire financial system, and the cost of them will be felt by European tax payers, notwithstanding that their benefits will affect even operators that are not incorporated in the European Union.[51]

We should consider the systemic effects of the option (to be taken by the ESFS) of forms of regulation that, at least in part, shall be realized through the imposition of restrictions to banks, financial intermediaries, insurance companies, and pension funds. This option, on the one hand, can be a valuable deterrent to take risks in sectors subject to public supervision. On the other hand, it introduces a clear asymmetry between the business conditions in Europe and in third-party countries.[52]

It should, however, be remembered that the mere extension of the scope of prudential regulation, to operators who are engaged in the shadow banking operations, involves a commitment of "own funds" to support the lending process. It does so by imposing a legal obligation on the part of persons belonging to only the European system.

It is clear that such a solution—while resolutely assessing the problems that characterize the present case—is not applicable without high transaction costs (of the own funds) and beyond the boundaries of the EU internal market. Therefore, it is difficult to exclude the possibility that in the future regulatory arbitrage will be realized that can trigger, in foreign markets, turmoil similar to that recorded in recent years. This applies, in particular, to the possibility that the supervision model in question is not adequate to safeguard the positive effects of the shadow credit intermediation process (in terms of increased liquidity) or to manage its negative effects (which are determined as a result of the risks that characterize the phenomenon).[53]

In other words, it appears to be necessary that the ESFS come to define a minimum standard of safety of the process, agreed with the industry. To conclude on this point, it should be noted that the provision of specific rules that lead shadow banking within the scope of the ESFS could achieve the goal of stability (that European organizations have set for themselves in the construction of the European Union), at a high cost (of capital).

8.8 The tasks of the European Banking Union

The book has now reached a point where the threads of the analysis can be gathered together. After dealing, in the previous paragraphs, with the recent trends in EU regulation of the internal market, one must take into account the review of the "financial apparatus" of the euro zone. In responding to the demands raised by the crisis, this review led to the creation of institutional changes with significant impact, intended to extend the scope of European supervision to affecting operations that until now could be fully traced back to the shadow banking system.

In particular, the consolidation of supervisory and monetary powers in the hands of one financial supervision (i.e. the ECB) will facilitate an action plan able to take into account the shadow entities (and their operations), and this in addition to the possible inclusion of the latter within the scope of the single supervisory mechanism.[54]

This does not refer only to the cases involving banks in the provision of services (e.g. loans and lines of credit, guarantees and sure ties, placement of securities and derivatives trading) and the transfer of assets (i.e. sales of receivables, securitizations, and other synthetic operations) to the shadow banking system. There is a more general perspective, which takes into account the financial market (as a whole) and the need for stability (as a requirement for the wellness of EU citizens).

The potential effect on the shadow banking system of the new supervisory mechanism may be appreciated. In particular, the overcoming of the structure that was determined from the early years of this century, when the option for a common monetary policy was accompanied by the choice of maintaining a national supervision (in pre-existing forms and, therefore, maintaining the division of the controls).[55]

It can also be said that (monetary and supervisory) policy integration, in the hands of the ECB, recalls the unity of the credit-money phenomenon, which had been experienced in some European countries (including Italy and Germany) during the twentieth century.[56] Putting this into perspective, it could increase the systemic gap between EMU countries and those that have not joined (including the United Kingdom).

So, within the EMU, the shadow banking system will have to remain inside its boundaries or conform itself to the new regulatory framework, intended to facilitate the completion of the European internal market and achieve the further aim to safeguard the stability (minimizing market failures and managing the competition with alternative systems). That is, also considering the integrated management of the mechanisms of supervision, deposit guarantee, and resolution of the crisis of credit institutions.[57]

On this point it should be noted that the supervisory tasks entrusted to the ECB by EU Regulation no. 1024/2013 appear likely to interact with those previously placed in the hands of the ESFS. That is, according to a principle of "close cooperation" that—as anticipated—will ensure an adequate level of supervision in the European Union (art. 4, Regulation 1024). Similarly, that also has to be said for the relationship of the ECB and the national authorities of the countries that have adopted the euro. This is also subject to the "duty to cooperate in good faith and the obligation to exchange information" (art. 6, Regulation 1024). In the case of other members of the European Union, however, "close cooperation" with the national authorities must be established by a special "decision" (art. 7, Regulation 1024).[58] Despite this, the regulatory framework is not intended to eliminate the conditions that, in recent years, led to the success of the shadow banking system.

From another perspective, it is important to come back to the diversity of approaches existing between the EMU area and the EU area. The latter, in fact, is impacted by "European decision-making processes," which are anchored to vetoes and qualified majorities that do not always make it possible to reach the levels of uniformity that qualify the euro zone. There are, therefore, specific design difficulties facing the leaders responsible for the European integration, as it is necessary to reach an equilibrium between the stability of the internal market, the coordination of financial supervision, and the efficiency of political governance in the whole of Europe.

It should also be considered that the supervision of the ECB will have a larger toolbox (than that of the ESFS), resulting from the possibility of combining the authoritative powers with the monetary ones, in order to safeguard the regular circulation of capital (and hopefully a sustainable future economic and financial growth).[59]

In conclusion on this point, I shall highlight that the asymmetries between the European Union and the EMU affect the implementation of an adequate system to supervise the shadow banking system, and call for a twin-track to the relative regulation process. Only close cooperation between the authorities of the ESFS and the ECB will ensure the establishment of uniform standards, able to avoid regulatory arbitrage favoring the proliferation of systemic risks.

That said, it can be mentioned that the regulation process of the shadow banking system will be able to take advantage of the "area of more intense cohesion" that—as mentioned earlier—corresponds to the euro zone. Indeed, if, at first, the distinguishing feature of this area had been the mere union of monetary policy, today the definition contains useful features of the legal order of the phenomenon under observation.[60]

In light of the foregoing, we can understand that the establishment of the single supervisory mechanism—and, therefore, the achievement of the EBU—cannot be the only answer to the problems that the shadow banking system raised during the initial phase of the financial crisis. It will also have to envisage a way out of the impasse that penalizes the start of a strong economic recovery. In other words, the unique mechanism of supervision will have to restore those techniques (bank based) that, for a long time, have ensured the financing of the industrial progress of Europe (and, thus, supported the sustainable development of the levels of wealth).

Consequently, the start of the activities necessary for the assumption of supervisory functions by the ECB—introducing new controls—achieves the objectives of transparency, safety, and confidence. Obviously, it must be taken into account that these activities, in addition to improving the quality of information available (and, in particular, that of the situation of systemic banks in the EMU), will have to identify the corrective measures necessary to ensure the safe and sound management of credit institutions, as well as the overall stability of the traditional banking system.

But, this will also affect individuals working in the shadow banking system, who—despite being outside the scope of supervision—will have to design operations compatible with the supervisory rules of the EBU. They will, therefore, also have to provide new techniques for involvement of banks (in order to continue to receive both loans to enter the production cycle and the banking services required).

This conclusion, moreover, is confirmed by the ECB's timely exercise of options—contained in art. 33, paragraphs 3 and 4 of EU Regulation 1024 of 2013. According to this regulation, in effect from November 3, 2013 the ECB can begin fulfilling the tasks assigned to it and may ask for all relevant information to be gathered through assessment of credit institutions.

It should be borne in mind that the results of the assessments undertaken by the ECB draw an uneven map of the banking industry, with significant areas of weakness in some markets. Important on this point are the results of the "stress tests" published on October 26, 2014, that highlight the conditions in which the national supervisory authorities deliver to the ECB the systemic banks in their country.[61]

With particular reference to Italy, the stress tests in question show a banking sector that—under a specific analysis—appears to be inadequate to the needs of the second largest manufacturing sector state in Europe.[62]

Hence, I shall take into account the prospect of a possible ambivalence of the interventional policies of the ECB as directed to take into account the national economic differences and, therefore, the supervisory practices adopted in the past by the competent authorities (I refer in particular to the forms of excessive flexibility shown towards the big banks during their capital strengthening).

So, in the new organizational system, the prevalence recognized by Regulations nos. 1022 and 1024 of 2013 to the functional profiles of the supervision highlights an innovative approach, pre-ordered to the homogenization of the banking market and, therefore, to avoid regulatory disparities that can undermine the smooth and uniform circulation of capital in Europe. In this perspective, the transfer of tasks to the ECB by national authorities corresponds to the need to avoid a *deduplement rationelle* (e.g. as could be the case of the access to the activity of credit institutions under art. 14 Regulation 1024). Therefore, the shadow banking system will have to deal with a system of "authoritative summit" that appears due to the associative type, sometimes in conjunction with functions attributed to a number of parties (ESRB, EBA, ESMA, and EIOPA national authorities). In front of such an organizational architecture, there are the warning signs of a "supranational community," organized in relation to the requirements of regulation and control that relate to the objectives of the EU.[63]

Ultimately, it can be said that the inclusive nature of the European project seems likely to reduce the scope of the financial market *fully deducted* from the public supervision. This approach appears to comply with the political intent of ensuring effectiveness of the project (of the European Commission) to adopt a set of rules in line with the standards promoted by global regulators in that field. Technique and policy, therefore, appear to be oriented towards the implementation of a program for stability and growth that takes into consideration also the alternative systems of circulation of wealth. It is clear that within this framework—*pre-ordered* to the correction of (market) failures and to the increase of transparency (of information)—the rules of supervision will have to be applied to the shadow banking system, even if in proportion.

8.9 The impact of the new targeted longer-term refinancing operations of the European Central Bank

New "monitoring requirements" derive from the opening of the ECB to the acceptance of ABSs as collateral, placed by the rules of the new "targeted longer-term refinancing operations."[64] I am not suggesting that these transactions recognize a particular legal status of certain financial instruments produced by the shadow credit intermediation process. I am saying, instead, that this "acceptance" causes an overture of the monetary policies in considering the results of financial innovation, and then the shadow operations. In other words, the calibration of the monetary policy shall have implications for the role that, to the present, the shadow banking system plays within the mechanisms of wealth distribution.

Of course, being in the presence of anti-crisis action (and not a structural reform of money market mechanisms), it should be noted that the monetary authority can achieve its purposes—of stability and growth—within the market through derogating actions of the ordinary course of business (in order to support the economic recovery, through the refinancing of loans granted to companies).[65]

For the purpose of this analysis, therefore, consider the possibility of acceptance, in the monetary policy of reference to the ABS or CDOs of the shadow banking system. In particular, there is the eventuality that these securities are in the assets of supervised entities (and, as such, qualified to entertain relationships with the ECB). This suggests a new form of interconnection between (i) regulated and non-regulated systems, (ii) banks and shadow banking entities, and (iii) banking and shadow banking operations.

In this case, there shall be obvious effects at macro-prudential level, even within the ECB, as the shadow banking risks may also affect the economy of the institution that accepts as collateral ABSs and CDOs issued at the end of the shadow credit intermediation process. It is this new interconnection that requires the public control of the quality of the financial instruments in question. This quality must be ensured both by the reliability of the underlying assets (i.e. the loans) and the correctness of the stages of the shadow credit intermediation process (i.e. the operations). The same has to be said for the verification of the conditions of information transparency, necessary to know the characteristics of the instruments in question and to measure their value correctly.

Naturally, the safety—of the monetary transactions guaranteed by ABSs and CDOs—requires the implementation of mechanisms for data gathering, as well as a monitoring system, and, ultimately, a new form of supervision. At this level, in fact, the aforementioned large gap that exists between the intensity of banking supervision and the lack of controls on the market-based circuits is not justifiable.

In conclusion, the recent monetary policy decisions present the EMU with the challenge of filling the significant gaps in the monitoring of relations that take place in the shadow banking system. This can affect the pursuit of monetary stability, which the ECB has placed at the base of the aforementioned TLTROs.[66]

8.10 Evolutionary trends of European supervision (following Directive 2014/65/EU)

In attributing a peculiar significance to the action of European authorities, one is called to deal with the recent indications that suggest a supervision of the shadow banking system anchored to the anti-crisis action (that has dominated also the activity of any global regulator in the last five years, at least). There is no doubt that there is persistent, directional change in the regulation adopted to prevent further degeneration of the financial markets, and—according to this pattern of evolutionary trends—the new supervision shall avoid the investment of the money in systems that produce pro-cyclical effects. This goal implies that attention should be paid to the shadow credit intermediation process, from which may originate an excessive amplification of the quantitative data (due to the aforementioned use of the fair market values).

The approach adopted by the regulators was summarized, in 2011, by the FSB as the possibility that "consolidation rules should ensure that any shadow banking entities that the bank sponsors are included on its balance sheet for prudential purposes." This led to the further consequence that "limits on the size and nature of a bank's exposures to shadow banking entities should be enhanced." This had the obvious effect that "the risk-based capital requirements for banks' exposures to shadow banking entities should be reviewed to ensure that such risks are adequately captured" (and this, both with regard to the granting of credit facilities to support—also implicitly—the shadow banking operations, and with reference to the investment in ABSs and CDOs).[67]

To this "starting point" are added, then, the instances for a review of the regulation of mutual funds and, in particular, of those that have an alternative nature (i.e. AIF). It was, in fact, noted that in the European market for alternative investment, fund managers (i.e. AIFM) administer substantial assets, contributing significantly to the determination of the volumes of trading that are running in the regulated trading venues.[68] It goes without saying that this alternative condition can help to spread or amplify risks through the financial market. Hence, the interconnection between the aforementioned funds and the shadow banking system makes it necessary to have a regulatory and supervisory framework that is "harmonized and strict" regarding management activities carried out within the European Union. These reasons will justify the adoption of Directive 2011/65/EC, which results in an improvement in the regulation of hedge funds and, therefore, a part of the shadow banking operations. This is clearly in line with the objective of ensuring a high level of investor protection (that, in this case, requires a common framework for the authorization and supervision of AIFM).[69]

A different set of considerations is associated with the need to redefine the economic incentives associated with the procedures of securitization. It should be noted that the European regulatory framework is geared towards the goal of ensuring the collection of statistical information by the ECB about what is necessary for accomplishment of the ESCB's tasks.[70] This leads to a call of duty for the arranger (of the relevant financial plan), which might require proceeding according to standardized and transparent methodologies, and subscribing to a part of the offering (in order to reduce the risk of moral hazard).[71]

From another perspective, with particular regard for the European financial markets, it is necessary to consider the construction made by Directive 2014/65/EU to satisfy the needs of the increasing number of investors and the growing range of services and instruments offered. That is, in relation the management of the new levels of complexity, and in search of better conditions of transparency and functioning of trading venues.

It seems that this regulatory intervention is oriented towards the contrast of the negative effects of the financial crisis. However, the arising rules are heading towards a further objective: the need to develop a single European rulebook applicable to all financial institutions in the internal market (as provided by Recital 6 Directive 2014/65/EU).

Consider, then, the fact that the directive aims at introducing "systematic regulation," governing the execution of any transaction in financial instruments, independently from the trading methods used. This is because such a circumstance ensures the supervision of the market-based financing transactions that might affect the integrity of the capital market.

Therefore, this regulatory framework shall be intended to cover also the new generation of organized trading systems that give content to the shadow banking, hence the need to ensure that such systems do not benefit from regulatory gaps (Recital no. 13, Directive 2014/65/EU). There is, however, not much to be said about the stage of implementation of the plans to strengthen financial supervision, set out in the conclusions of the European Council in June 2009. That is, given that the abovementioned Directive no. 2014/65/EU is a step forward towards a new "level playing field" that can increase investor protection, but not the "season finale."[72]

Nothing, in fact, prevents the need for further interventions in the markets for financial instruments. The European Commission is aware of that and, even before the publication of the Directive 2014/65/EU, proposed a regulation on the reporting and transparency of securities financing transactions.[73] Again, the primary objective of this proposal is to eliminate the gray areas that have a potential impact on systemic risks (and, consequently, extend the regulation and oversight to all financial institutions, instruments, and systemically important markets).

In particular, it is imperative to deal with the possible (or, rather, proposed) provision of monitoring procedures for systemic risks, which are designed to measure the accumulation of the latter (due to securities financing transactions). It is evident how, with these procedures, it is desired to allow the dissemination of information (and, thus, also the contractual transparency of re-commitment activities); that is, in order to implement forms of conscious investment.

There are measures in place that aim to regulate the successful move of a substantial part of intermediary activities beyond the boundaries of the traditional banking sector. To the effectiveness of these measures is, therefore, connected the possibility of limiting the opacity of the financial system, in order to perceive and correctly price the risks of the case.

Conversely, there is no trace of proposals wanting to prohibit this form of financing. Perhaps it isn't time to deprive the financial operators of tools that could lead to the creation of credit (through maturity transformation) and increased liquidity (although the same resolve in leveraging).

In conclusion, there is clearly a general regulatory trend towards increasing levels of transparency (which shall be guaranteed by appropriate disclosure requirements and effective means for the circulation of information). More generally, the regulators seem to be oriented towards the conservation of the circuits of shadow banking, but they want to prevent the alternative from undermining investor confidence and, in that way, prevent the recovery of the necessary conditions of stability and growth for the smooth functioning of markets.

8.11 The role of global regulators: the World Bank and the International Monetary Fund

Doubts have been cast in this chapter that a systemic oversight on the shadow banking system can be incorporated only at global level, having (the supervisor) to provide a set of incentives and safeguards that must control the capital circulating outside the banking industry. As was anticipated, there is more than one institution able to deal with the supranational dimension of the operators, and the cross-border nature of the transactions that give content to the shadow credit intermediation process.

Certainly the absence of a global legal system responsible for the care of the economic and financial interests that occur on a global level is an obstacle to the affirmation of a supervisory system.[74] Therefore, one can only assume a two-phase adjustment process, in which alternate inputs come from the international economic institutions (lacking of authoritative powers: the World Bank, the IMF, and the FSB) and regulatory actions from nations armed with the necessary sovereign powers. In other words, at present, there is affirmation that a mechanism that plays an initial driving role in the drafting of legislation (with the single value of soft law) will then be incorporated—by the latter—in hard law.[75]

It goes without saying that, at international level, there is only soft law based on the statements made by the aforementioned subjects, which can *lightly* influence the phenomenon.[76]

Therefore, we shall consider the actions of the World Bank, which focused on the shadow banking operations that—as we have seen—can increase the levels of risk; not only with regard to the excessive use of leverage (and related pro-cyclical effects), but also with respect to the hypothesis of a shift to regular banking.[77] These are, of course, actions intended to recover the stability and the rate of growth of the global economy (that should be able to sustain the levels of wealth achieved in the twentieth century).[78]

I shall also highlight that the World Bank has played a key role in the containment of the most severe stages of the recent financial crisis, acting, in some respects, as "global economic policymaker."[79] This was achieved not only by anchoring its action on the type of instruments and the amount of resources available, but also by guiding the search for formal-legal parameters that connect the rehabilitation and rescue the economy from time to time.

In particular, the World Bank identified the adverse effects that occur in the "emerging markets and developing economies" (i.e. EMDEs). It detects, then, the attempt to measure the development of the phenomenon under observation, noting that "the sector is particularly large in the Philippines and Thailand (more than one-third of total financial system assets) and its share has been gradually rising."[80] That said, it should be noted that the World Bank has addressed the issue of shadow banking monitoring certain risks (which originate in this system), and wants to extend the reporting requirement, in order to improve the processes of data collection.

Despite the simplicity of the individual operations (in line with the backwardness of the financial conditions faced by emerging countries), the analysis of the conditions of the shadow banking system found in China also seem worthy of appreciation.[81] It was, in fact, recognized that in the latter the processes are not necessarily realized through the use of leverage or maturity transformation, but with the help of a wide range of entities, all of which are difficult to track down.[82]

Hence the proposal, by the World Bank, of a supervision system with the objective of preventing risks for financial stability, maintaining the conditions for the "shadow banks (to) play an important role in channeling alternative funding sources to the EMDEs, especially as the significant deleveraging pressure from European banks continues."[83] This means that, in a legal perspective, the *inputs* from the World Bank contribute to the development of a supervisory system that increases the security of shadow banking (as an alternative channel) and keeps its efficiency as well as the ability to finance the economies of emerging countries.

Consider now the intervention of the IMF in the shadow banking system, given the empirical evidence that *this system* interacts with the particular role (played by the IMF) in promoting monetary co-operation and protecting the stability of the exchange rates (especially during the recent financial crisis). The IMF, in doing so, extended—as well as regulated markets—its function of "economic surveillance," through which it keeps track of the economic health of its member countries, alerting them to risks on the horizon and providing policy advice.[84]

As a result of this intervention (and based on the analysis of global financial flows), the IMF has developed a "mapping" of the shadow banking system.[85] However, this research has not reached any final conclusion, and it has not clarified the effects of the "non-bank–bank nexus" that takes place in the shadow.[86] It is possible only to understand certain patterns of the growth of this system, given that the IMF suggests "a search for yield, regulatory arbitrage, and complementarities with the rest of the financial system play[s] a role in the growth of shadow banking."[87] Hence, where banks are apparently withdrawing from certain risky activities (in response to strengthened regulations), there is an opportunity to develop the aforementioned market-based credit intermediation techniques.

It shall be highlighted here that the IMF aims at structuring a monitoring system, rather than suggesting operational constraints of the shadow banking entities. Indeed, this methodology seems appropriate to manage the consequences of the regulatory options of the Dodd-Frank Act and of the Basel III Agreement, taking into account the transactions that—for their risk—are placed outside the banking sector.[88] In addition, this approach directly addresses the regulation of the shadow banking risks, calling for an early adoption of the program of action suggested by the FSB.

In light of the foregoing, it is clear that the IMF and the World Bank, as much as they are oriented towards complementary aspects of the matter in question, have aroused the interest of the international community in the phenomenon under observation. Since the beginning of its analysis, the IMF has focused on the macroeconomic aspects and financial interconnections, and the World Bank on the long-term effects resulting from the development of a system of credit intermediation alternative to the traditional one.[89]

In addition, it should be noted that the cooperation of these bodies with the G20 industrialized and emerging market economies happens with a view to strengthening the supervision on the system in question and achieving a broader reactive process aimed at containing specific adverse effects (and preventing further degeneration of economics and finance). Hence, there is explicit confirmation of the technical nature, due to the construction of the forms of supervision on the shadow banking system. The orientation towards the objectives of stability and growth can also be deduced, rather than towards social purposes (which traditionally characterize the economic constitutions of democratic countries).

There is, therefore, open a space for discussion and debate, in which the policy makers will have to check the social value of the rules proposed by global regulators and, if affirmative, transpose the contents in a disciplinary action consistent with all of the rights at the basis of the community.[90]

8.12 Limited effects of the supervision designed by the Financial Stability Board

The conclusion is, obviously, that the most significant inputs for the construction of a system of regulation and control on the shadows are those issued by the FSB. I shall not argue the option for a "two-step process" in which "first, for monitoring purposes, authorities should cast the net wide, looking at all non-bank credit intermediation to ensure that data gathering and surveillance cover all areas where shadow banking-related risks might potentially arise. Second, authorities should narrow the focus for policy purposes to the subset of non-bank credit intermediation involving maturity/liquidity transformation, imperfect credit risk transfer, leverage, and/or regulatory arbitrage concerns."[91]

However, in some other respects, the foregoing is moderately conservative in its implications. This approach appears to reflect the importance attributed by the international financial community to the increasing involvement of supervised entities to unregulated banking activities. The FSB, in fact, was quick to note that the phenomenon in question had become—before the crisis of 2007—the circuit preferred by professionals (banking and non-banking) to finance each other, as well as to enable speculative transactions (through the activities of duration, maturity, and liquidity transformation).[92]

Nevertheless, one should agree with the FSB's decision not to proceed with the contrast of alternative circuits of capital, but to initiate an intervention aimed at safeguarding this form of intermediation (despite the possibility that it would become a source of risk of a new type, interconnected with the banking systems and, therefore, dangerous to the stability of traditional markets). There are also clear reasons to agree with the promotion, by the FSB, of a regulatory system that, in pursuing its "tutioristic objectives," aims at ensuring a complete protection of the market, taking into account the broader areas followed by those limited by the legal boundaries imposed by the current supervisory regime. Obviously, only the European regulator is able to ensure the highest levels of safety, by transforming the inputs of the FSB in hard (European) law. This is one of the most likeable paths to implementation of the "high-level principles for monitoring the shadow banking system" proposed by the FSB.[93]

This is, therefore, a proposal that will flow into several workstreams, which will be consolidated into one piece of legislation that can uniquely regulate entities, operations, and the risks associated with both. Therefore, the abovementioned inputs have to be completed with regard to the empirical findings of the mapping process, in order to draw a comprehensive legal framework (that is also commensurate with the actual needs of the financial market).

Moreover, the aforementioned approach requires the identification of the transactions that, within the shadows, feed the "key systemic risk factors." It is evident that even the inputs in question should be incorporated in hard laws, in order to avoid market failures. This applies, in particular, to the need to include specific limits on the subject, in order to reduce the gap between the maturity of the "short-term liabilities" and the one originally planned for the initial loans and other "long-term assets" that make up the estate of the shadow banking entities. (Also in order to measure the proper levels of market liquidity of ABSs, CDOs, and business operators, to avoid insolvency of the shadow banking operations.)[94]

Essential, in the construction of the supervision system suggested by the FSB, will be the management of the dynamics of "credit risk transfer," to which corresponds the need to adapt the rules of the balance sheets to the new effects of the shadow banking operations (which are not always properly recorded in the accounting international principles).[95] Therefore, it seems appropriate here to refer to prudential filters that can measure the "off-balance sheet exposures" of the banking industry (and other forms of risk taking implied).

I shall consider, on this point, the proposal of the FSB to introduce an obligation for the measurement of risks, and their actual exposure in reference to the quality of the mitigation techniques (and other forms of coverage) that each subject can activate in the relevant market. It goes without saying that, in the presence of such a system, prudential measures must take into account the ability to manage, transfer, or sterilize risks, by putting in the hands of the supervisory authorities the task of monitoring the circulation of the latter and assessing the overall resilience of the system.[96]

Finally, the focus will turn to the inputs designed to reduce the leverage, which are summed up in a quantitative limit on indebtedness of a shadow banking entity, or in the control of multiplier effects related to the "daisy-chains" that recur in this case. This seems to be in line with the information collected for computing the leverage ratio under the Basel III framework, but without providing an uncritical application of the Basel Agreement to the shadow banking system. However, there has already been occasion to observe that the extension of the prudential supervision to the system under observation would have negative effects. That is, both due to the high cost of the latter and the reduction of the competitive advantages of the shadow credit intermediation process.[97]

That said, it should be noted that the FSB is oriented towards the construction of a uniform system of supervision that shall be able to internalize, in shadow banking, the risks of transactions. Nevertheless, there is no *strong evidence* that the inputs reach the higher prudential standards; but it does seem that they try to keep the benefits of the shadow. This is not free of positive externalities, because "low transaction costs" shall facilitate the financing of the real economy at more convenient rates than those offered by banks in the context of prudential supervision.[98]

It is clear that the ultimate goal of the FSB cannot be the repression of any attempt to abuse (the possibilities offered by the shadow banking system), despite the provision of specific measures to prevent speculation or regulatory arbitrage. It is preferable to suggest a "monitoring process [that] should be sufficiently flexible, forward-looking and adaptable to identify new shadow banking activities, and to capture important innovations and mutations in the financial system."[99]

In light of the foregoing, the approach to the regulation of the shadow banking system seems to be part of the general orientation of the FSB towards the strengthening of public control in the capital market. If this approach promotes the expansion of the areas of traditional matrix supervision (and consequently tries to reduce the set of operations that can be legitimately realized without the government oversight knowing about it), it is also true that it does not provide urgent (and expensive) mechanisms of oversight.[100]

In conclusion, the importance of the *inputs* formulated by global regulators and the essentiality of the implementing *actions* of the national authorities should be understood. If the former appear capable of managing the size of the phenomenon, the latter are required to establish effectiveness (or enforceability) to the precepts adopted to guide the dynamics of the shadow banking system towards the sound management of the available resources. But the task of increasing the transparency in the financial market is still in the hands of the (regional or national) legislative authorities; in the hope that they will not create asymmetries between local markets subject to different legal order (and in particular between the United States and the European Union, or between the latter and the EMU).

Conclusions

The shadow banking system seems to be the outcome of the deregulation process begun in the twentieth century. Even if the latter was aimed at reanimating the real growth of the economy and speeding up the circulation of money, the underlying idea was that a highly regulated financial market slowed down business and decreased the wellness of society.

Probably, at the beginning of the third millennium, the trends toward easy credit and, therefore, the development of risky financial activities were only the natural consequence of such an idea. This also explains the opening of domestic supervisors to cross-border activities as another attempt to increase volume and competition, according to the will to maximize the return of the activities performed in the new deregulated market. Even in this case, perhaps, the lack of transparency and the regulatory arbitrage were only another consequence of a development policy. Undoubtedly, the devolpment of financial innovation was another step forward toward the goal of "maximum production," aiming at capitalizing any transaction that occurs in the markets. Undoubtedly, the lack of resilience was only one of the degenerative elements of an advanced capitalism, which shows its limits when forced.

These are the main features that, in the book's analysis, explain the growth of the shadow banking system and the decline of financial regulation, until the capital markets started dealing with the crisis of 2007 (and its consequences).

This is why the book moved from regulatory evidences showing how the market for credit, before reaching its maximum, splits into two channels: the traditional and the shadow banking system. The first should be the safer system for collecting savings; the second should provide credit to the real economy without the prudential limits or the monetary safeguards. Hence, the safety costs, and the cost for protection, would have reduced the possibility of financing the real economy. Sharing this evidence, the book has explained the limited scope of banking supervision, and—with it—the high capital requirements introduced by the regulation adopted in Basel. But there is no justification for the lack of transparency, the information asymmetries, and—more generally—the absence of a system of data management.

These questions have been addressed throughout the book, and—while dealing with the turbulences, the effects of the anti-crisis measure, and the proposal of regulation of international authorities—I reach the conclusion that this alternative channel has a high level of opacity. In other words, it is understood that the shadow credit intermediation process—in the absence of a credit institution—amplified the need for "data and news." This is true for both the investment choices and the forecasts on market trends.

Definitely, the regulatory intervention must increase the transparency in order to prevent, in these shadows, the operators pursuing unlawful intents. The current acts of the European Union aim in this direction, showing the strengthening of the credibility of authorities' commitments to implementing a new system of oversight that is able to prevent any crisis of any entity before it jeopardizes the markets in which it operates.

The introduction of a "banking union" and the "capital market union" perspective show alternative ways of linking investors and savers with growth. Understanding this, it can be realized how the European Commission pursued its initiatives to create a safer and sounder financial sector for the single market, renewing the prudential requirements for banks, improving depositor protection schemes and rules for managing failing banks. We can also appreciate the deeper integration of the banking system of those member states that share the euro, and then the establishment of a single supervisory mechanism and a single resolution mechanism for banks.

If it is expected that the shadow banking system will receive a part of the ECB expanded asset purchase programs, then it is possible to hypothesize new forms of supervision for this intermediation channel (when the latter recives the resources of LTROs, TLTROs and quantitative easing program). In this context, EU plans seem to be in line with the recent decision to promote new control devices, carried out by the FSB in its 2014 Report, in order to avoid the default of a single operation extending its negative effects to the whole shadow process of credit transformation.

As has been seen, the most interesting regulatory proposal goes beyond the redefinition of the scope of the capital adequacy parameters required by Basel III (and, in particular, of the funding ratio), being projected towards an action that is able to create a sustainable market-based finance industry for alternative credit support to real economy.

In conclusion, the remarks on the strenghting of the supervision over the shadow banking system shall be in line with an interdisciplinary approach used by public authorithies during the crisis. The expectation, then, is that the regulatory process will pursue both the economic efficiency and the equal distribution of the resources.

On this point, the analysis ends by taking into accunt the current goals of the global regulators, which choose to promote a new common international standard on total loss-absorbing capacity (for global systemic banks). It seems, therefore, possible that the monetary authorities shall use their *powers* to manage the expansion of the shadow banking system. The possibility that the ECB and other supervisory authorithies will not allow banks to invest depositors' savings in the shadow credit intermediation process should not be disregarded, but this—obviously—implies an authoritative intervention that cannot be based only on regulation and control, being necessary to realign market incentives to turn the demand for loans toward banking and financial intermediaries, which fall within the scope of the public supervision (and benefit from the related guarantees).

Finally, focusing on the European context, the will to increase the transparency of the financial market calls for a *political European Union* that goes far beyond the current structure of the banking and capital markt union. As Winston Churchill's "Speech to the academic youth," at the University of Zurich in 1946, concludes: "there is a remedy which... would in a few years make all Europe... free and... happy. It is to re-create the European family, or as much of it as we can, and to provide it with a structure under which it can dwell in peace, in safety and in freedom. We must build a kind of United States of Europe."

Notes

Introduction

1. For further information we must take into account that the first attempt to map the shadow banking system was published by the FSB as part of the 2011 report, using data from eleven jurisdictions and the euro area. Then, this approach evolved continually in the following years.
   The 2014 report presents the results of the fourth annual monitoring exercise using end-2013 data, following the approach set out in the FSB report to the G20 in October 2011. This report includes data from twenty-five jurisdictions and the euro area as a whole, bringing the coverage of the monitoring exercise to about 80 per cent of global GDP and 90 per cent of global financial system assets; see FSB (2014) *Global Shadow Banking Monitoring Report 2014*, 30 October, pp. 1 and 38 where there are the results of *IMF"s Global Financial Stability Review*, October 2014, *statistical appendix*.
   It is helpful to refer to Williamson (1985), *The Economic Institution of Capitalism* (New York), p. 385 ff. for a digression on risk neutrality, which helps in understanding the role of the law under a perspective that emphasizes the "transaction costs" and, so, is useful for the purposes of this analysis.
   From a regulatory perspective, see Capriglione (2010), *Misure anticrisi tra regole di mercato e sviluppo sostenibile* (Torino) p. 43 ff. on the role of the global regulatory network in the financial markets.
2. See Pilkington (2008) "Conceptualizing the Shadow Financial System in a Stock-Flow Consistent Framework," *Global Business & Economics Anthology*, Vol. 2, p. 268 ff. where there is a first definition of a "shadow financial system," started with the growing of an institutional sector composed of all the unregulated non-banking financial institutions (NBFIs) that fall outside the regulatory scope of central banks
   See also the FSB (2014) *Global Shadow Banking Monitoring Report 2014*, October 30, where it is summarized that "The shadow banking system can broadly be described as credit intermediation involving entities and activities outside of the regular banking system. Intermediating credit through non-bank channels can have important advantages and contributes to the financing of the real economy; but such channels can also become a source of systemic risk, especially when they are structured to perform bank-like functions (e.g. maturity and liquidity transformation, and leverage) and when their interconnectedness with the regular banking system is strong. Therefore, appropriate monitoring of shadow banking helps to mitigate the build-up of such systemic risks."
3. See Schneider-Williams (2013) "The Shadow Economy", *Institute of Economic Affairs Monograph, Hobart Paper*, no. 172, who suggest that the main drivers of the shadow economy should be: tax and social security burdens, tax morale, the quality of state institutions and labour market regulation. This is why

the authors highlight that a reduction in the tax burden is, therefore, likely to lead to a reduction in the size of the shadow economy.

For the pre-crisis approach to this topic, see Davidoff and Solomon (2007), "Black Market Capital", *Wayne State University Law School Research Paper*, no. 07-26, where the author uses the abovementioned terms to indicate certain supply and demand, which "drives public investors to substitute less-suitable, publicly available investments which attempt to mimic the characteristics of hedge funds or private equity." It is also highlighted that, in the United States, "Yet, current federal securities regulation effectively prohibits the public offer and purchase in the United States of these hedge fund and private equity investments. Public investors, foreclosed from purchasing hedge funds and private equity, instead seek to replicate their benefits."

See also Fardmanesh-Douglas (2003) "Foreign Exchange Controls, Fiscal and Monetary Policy, and the Black Market Premium," *Yale University Economic Growth Center Discussion Paper*, no. 876, for the standards of examination of the relationship between the official and parallel exchange rates used before the crisis (in the paper, the authors take into account the data of three Caribbean countries, Guyana, Jamaica, and Trinidad, during the 1985–1993 period using co-integration).

4. See Ostrom (2005) *Understanding Institutional Diversity* (Princeton), p. 135 ff. for an analysis on the sense of permission rules, and of how the institutions are formed, operated, and influence behavior in the markets.
5. See Tyson and Shabani (2013) "Sizing the European Shadow Banking System: A New Methodology," *CITYPERC Working Paper Series*, no. 2013/01.
6. On this point, we must refer to the statement of the Federal Reserve Bank of New York about the fact that "Credit intermediation involves credit, maturity, and liquidity transformation. Credit transformation refers to the enhancement of the credit quality of debt issued by the intermediary through the use of priority of claims.... Maturity transformation refers to the use of short-term deposits to fund long-term loans, which creates liquidity for the saver but exposes the intermediary to rollover and duration risks. Liquidity transformation refers to the use of liquid instruments to fund illiquid assets"; see Pozsar, Adrian, Ashcraft, and Boesky (2012) *Federal Reserve Bank of New York Staff Reports – Shadow Banking*, no. 458, July 2010 (revised February 2012), p. 5
7. See Sakurai and Uchida (2013) "Rehypothecation Dilemma: Impact of Collateral Rehypothecation on Derivative Prices Under Bilateral Counterparty Credit Risk", *25th Australasian Finance and Banking Conference 2012*, for the background of our analysis, given the practice where a derivatives dealer reuses collateral posted from its end user in OTC derivatives markets. From this and other speculative practices arises the need for regulation that will be investigated in Chapter 6.

See also Stiglitz (2012) *The Price of Inequality* (New York), p. xi, on the specific problems raised by the failure of markets in the recent crisis, and p. 52 for the implications on inequalities in the financial markets.
8. See Skeel (2010) "The New Financial Deal: Understanding the Dodd-Frank Act and its (Unintended) Consequences," *U of Penn, Inst for Law & Econ Research Paper*, no. 10-21, where it is clearly explained why the US regulation

has the goals of limiting the risk of the shadow banking system by more carefully regulating derivatives.
9. See Luck and Schempp (2014) "Banks, Shadow Banking, and Fragility", *ECB Working Paper*, no. 1726, where the authors highlights that the relative size of the shadow banking sector determines the stability of the financial system.

See also Adrian and Liang (2014) "Monetary Policy, Financial Conditions, and Financial Stability," *FRB of New York Staff Report*, no. 690, where the authors review monetary policy transmission channels, focusing on vulnerabilities that affect monetary policy's risk–return trade-off, including 1) pricing of risk, 2) leverage, 3) maturity and liquidity mismatch, and 4) interconnectedness and complexity.

We must take into account the approach of Troeger (2014) "How Special Are They? – Targeting Systemic Risk by Regulating Shadow Banking," *SAFE Working Paper*, no. 68,where the relevance of the phenomenon is measured by the financial stability concerns associated with shadow banking.

In particular, see Sharma (2014) "Shadow Banking, Chinese Style," *Economic Affairs*, Vol. 34, Issue 3, p. 340 ff. where he describes the phenomenon as "poorly regulated, engaging in opaque forms of intermediation, deeply interconnected with the official banking system, and operating with implicit government guarantees, they pose a major source of systemic risk." On this basis, the author tries to explain the rapid proliferation of shadow banks in China.
10. See Kim (2014) "Money is Rights in Rem: A Note on the Nature of Money," *Journal of Economic Issues*, Vol. 48, Issue 4, pp. 1005–1019, where the author aims to demonstrate that contemporary banking, including commercial and shadow banking, creates money by mirroring credit in the image of rights in rem.
11. See Rixen (2013) "Why Reregulation after the Crisis is Feeble: Shadow Banking, Offshore Financial Centers and Jurisdictional Competition," *Regulation & Governance*, Vol. 7, Issue 4, p. 435 ff. where the author observes incremental and ineffective measures instead of a swift and radical regulatory reform in that sector after the crisis. This is why it is highlighted that, at the international level, governments are engaged in jurisdictional competition for financial activity, while, at the domestic level, governments are prone to capture by financial interest groups, but also susceptible to demands for stricter regulation by the electorate.
12. This is a fundamental criterion for contemporary legal orders, defined as "the benefits that individuals obtain from acts minus the harm done and the costs of enforcement of law," see Shavell (2004) *Foundations of Economic Analysis of Law*, (Cambridge), p. 575.
13. Please note that "the use of the term 'shadow banking' is not intended to cast a pejorative tone on this system of credit intermediation"; see FSB (2013) *Global Shadow Banking Monitoring Report 2013*, November 14, cit., p. 1.

See, in general, Krugman (2012) *End This Depression Now!* (New York) for a useful review of the "economic slump" that has afflicted the United States, the European Union, and many other countries in the recent years; see also Posner (2009) *A Failure of Capitalism* (Cambridge), p. 41 ff. on the banking crisis of 2008 and the descent into depression.
14. See *The G20 Seoul Summit Leaders Declaration*, November 11–12, 2010, p. 2.

15. See *The G20 Cannes Summit Commitments*, November 3–4, 2011.
16. See *The G20 Los Cabos Summit Leaders Declaration*, June 18–19, 2012, p. 6.
17. See Capriglione and Troisi (2014) *L'ordinamento finanziario europeo dopo la crisi* (Torino), p. 25 ff. on the limits of the politics in regulating the global financial system.
18. As 1984 Nobel Prize-winning economist J. Tobin (1984) observed in his article "On the Efficiency of the Financial System," *Lloyds Bank Review*, July, p. 1 ff.
19. See *Scientific Background on the Sveriges Riksbank Prize in Economic Sciences in Memory of Alfred Nobel 2013*, "Understanding Asset Price," compiled by the Economic Sciences Prize Committee of the Royal Swedish Academy of Sciences, October 14, 2013, p. 1.
20. See *Scientific Background on the Sveriges Riksbank Prize in Economic Sciences in Memory of Alfred Nobel 2013*, "Understanding Asset Price", cit., p. 3.
21. See, once more, *Scientific Background on the Sveriges Riksbank Prize in Economic Sciences in Memory of Alfred Nobel 2013*, "Understanding Asset Price," cit., p. 1, where it is highlighted that "if it is possible to predict with a high degree of certainty that one asset will increase more in value than another one, there is money to be made."

1 General Observations

1. In other words there is not a "Grundnorm"; see Kelsen (1954) *General Theory of the Law and the State* (Cambridge—Milano), p. 116 ff.; Hart (1961) *The Concept of Law* (tr. Cattaneo, ed. Torino, 1965), p. 126.
2. See *Recital no. 69*, Directive 2014/65/EU.
3. See Masera (2009) "La crisi globale: finanza, regolazione e vigilanza alla luce del rapporto de Larosiere," *Rivista Trimestrale di Diritto dell'economia*, I, p. 147; from a legal point of view, see Capriglione (2012) "Commento sub art. 5 d. lgs. 385/1993," in Capriglione (ed.), *Commentario al testo unico delle leggi in materia bancaria e creditizia* (Padova), Vol. 1, p. 49 ff. where the author clarifies the evolution of the banking supervision, from the Italian experience of the twentieth century to the ESFS of the new millennium; see also Capriglione (2013) *L'Unione Bancaria Europea* (Torino), p. 69 ff., on the new role of the European Central Bank after the Regulation (EU) no. 1024 of 2013.

   See also Sciarrone, Alibrandi, and Santoro (2010) "La nuova disciplina dei servizi di pagamento dopo il recepimento della direttiva, 2007/64/CE (D. Lgs. 27 gennaio 2010, no. 11)," *Banca borsa e titoli di credito*, I, p. 377 ff., on the new EU regulation of the Single European Payments Area and its legal effects on the internal market.
4. See Rixen (2013) "Why Reregulation after the Crisis is Feeble: Shadow Banking, Offshore Financial Centers and Jurisdictional Competition," cit., where there is an explanation of the regulatory initiatives on shadow banks and OFCs at the international level and within the USA and the European Union, where I focus on France, Germany, and the United Kingdom.
5. Please note that "the use of the term *shadow banking* is not intended to cast a pejorative tone on this system of credit intermediation"; see FSB, *Global Shadow Banking Monitoring Report 2013*, November 14, 2013, p. 1.

6. The Federal Reserve Bank of New York gave a brief description of traditional banking industry useful for the purposes of this analysis. They explained that "in the traditional banking system, intermediation between savers and borrowers occurs in a single entity. Savers entrust their savings to banks in the form of deposits, which banks use to fund the extension of loans to borrowers. Savers furthermore own the equity and debt issuance of the banks. Relative to direct lending (that is, savers lending directly to borrowers), credit intermediation provides savers with information and risk economies of scale by reducing the costs involved in screening and monitoring borrowers and by facilitating investments in a more diverse loan portfolio"; See Pozsar, Adrian, Ashcraft, and Boesky (2012) *Federal Reserve Bank of New York Staff Reports – Shadow Banking*, no. 458, July 2010 revised February 2012, p. 4.
7. See FSB, *Global Shadow Banking Monitoring Report 2014*, cit., p. 35 ff.
8. The reader should be aware that the term "shadows" will mean both the "shadow banking system" and the whole alternative financial market. As a poet suggested, "A people always end by resembling its shadow," Rudyard Kipling, quoted in Maurois, *The Art of Writing*, "The Writer's Craft," sct. 2, 1960.
9. To clarify this sentence, it is useful to highlight the analysis made by the FSB, given that "The first attempt to map the shadow banking system was published by the FSB as part of the 2011 report, using data from eleven jurisdictions and the euro area. The approach evolved continually in the following years. The 2012 report expanded the coverage to 25 jurisdictions and the euro area as a whole, while in this latest report, the granularity of data collected has been enhanced to allow for a refinement of the estimate of the shadow banking system. More specifically, the 2013 monitoring report presents some preliminary steps to narrow down the estimated size of the shadow banking system by filtering out non-bank entities and activities that do not pose bank-like risks to financial stability"; see FSB (2013) *Global Shadow Banking Monitoring Report 2013*, cit., pp. 1–5, where they analyze the results published in IMF's *Global Financial Stability Review*, October 2013, statistical appendix.
10. This explains the option to delegate these analyses to the Analytical Group on Vulnerabilities (AGV), and to the Technical working group of the Standing Committee on Assessment of Vulnerabilities (SCAV); see FSB, *Global Shadow Banking Monitoring Report 2013*, cit., p. 5.
11. See FSB, *Strengthening Oversight and Regulation of Shadow Banking Policy Framework for Strengthening Oversight and Regulation of Shadow Banking Entities*, August 29, 2013, p. i.
12. See Kodres (2013) "What is Shadow Banking?", *Finance & Development*, Vol. 50, Issue 2, available at http://www.imf.org.
13. See Claessens, Pozsar, Ratnovski, and Singh (2012) *Shadow Banking: Economics and Policy*, edited by IMF Research Department, December 4, p. 3, where it is stated that "these functions can be economically useful, and need to be understood and properly regulated."
14. See G20 (2014) *Communiqué*, Meeting of G20 Finance Ministers and Central Bank Governors Washington D.C., April 10–11.

15. See "The Shadow Banking System in the Euro Area: Overview and Monetary Policy Implications," in *Deutsche Bundesbank Monthly Report*, March 2014, p. 15 ff., where it is also stated that "so far, the Euro-system's monetary analysis has addressed the risk that increased shadow banking activity could reduce the information content of monetary indicators by incorporating selected shadow banking entities (money market funds) into the calculation of monetary aggregates and by adjusting these aggregates for certain transactions (e.g. securitization). These corrections – in combination with a more detailed analysis of sectorial shifts in money holdings – currently ensure that the data on monetary aggregates are sufficiently meaningful."
16. See Luck and Schempp (2014) "Banks, Shadow Banking, and Fragility," cit., for a banking model of maturity transformation. It is an interesting conclusion, according to which, if the shadow banking sector is small relative to the capacity of secondary markets for shadow banks' assets, shadow banking is stable.
17. See Meeks, Nelson, and Alessandri (2013) "Shadow Banks and Macroeconomic Instability," (Banca d'Italia, Roma), no. 939, November 2013, p. 3.
18. See Tucker (2012) *Shadow Banking: Thoughts for a Possible Policy Agenda*, speech of the Deputy Governor Financial Stability, Member of the Monetary Policy Committee and Member of the Financial Policy Committee, at the European Commission High Level Conference, Brussels, April 27, 2012.
19. I refer to the following definition of "instruments, structures, firms or markets which, alone or in combination, replicate, to a greater or lesser degree, the core features of commercial banks: monetary or liquidity services, maturity mismatch and leverage"; see Tucker (2010) *Shadow Banking, Financing Markets and Financial Stability*, January.
20. See Tucker (2012) *Shadow Banking: Thoughts for a Possible Policy Agenda*, cit., where it is highlighted that "for other shadow banks, liquidity services are offered without such back-up lines. In those cases, claims on the shadow bank have, in effect, become a monetary asset. Examples probably include money market mutual funds and an element of the prime brokerage services offered by securities dealers to levered funds." So, in this case, the shadow banking system does not proceed to the exercise of the credit, but uses the savings in market transactions.
21. See *Credit and Liquidity Programs and the Balance Sheet*, available at http://www.federalreserve.gov, October 30, 2014. I refer to eligible collateral for this program (and a certain number of other lending facilities) shows how—at least until February 2010 (within the AMLF program)—ABCP issuers and Money Market Mutual Fund received liquidity facility from the FED. Moreover, the effects of "Term Asset-Backed Securities Loan Facility Program," which shows how the Federal Reserve responded aggressively to the financial crisis, must also be taken into account.
22. See European Commission (2014) "Proposal on Transparency of Securities Financing Transactions," January 29, where there are the responses to the *Communication on Shadow Banking and Proposal on Money Market Funds*, September 4, 2013; the *Green Paper on Shadow Banking*, March 19, 2012; and the *Conference: Towards a Better Regulation of the Shadow Banking System*, April 27, 2012.
23. See European Commission, *Green Paper on Shadow Banking*, cit., p. 2.

24. See European Commission, *Green Paper on Shadow Banking*, cit., pp. 3–4.
25. See European Commission, *Communication from the Commission to the Council and the European Parliament, Shadow Banking – Addressing New Sources of Risk in the Financial Sector*, Brussels, September 4, 2013, COM(2013) 614 final.
26. It is not a coincidence that the Green Paper stated that "The FSB's work has highlighted that the disorderly failure of shadow bank entities can carry systemic risk, both directly and through their interconnectedness with the regular banking system. The FSB has also suggested that as long as such activities and entities remain subject to a lower level of regulation and supervision than the rest of the financial sector, reinforced banking regulation could drive a substantial part of banking activities beyond the boundaries of traditional banking and towards shadow banking"; see European Commission, *Green Paper on Shadow Banking*, cit., p. 2. Obviously, this is a statement which performs its duties as a piece of "soft law," given the absence of any regulatory power to enforce it.
27. See *Recital* no. 3, Directive 2014/65/EU.
28. See art. 2.1, Regulation (EU) no. 648/2012.
29. See *Recital* no. 13, Directive 2014/65/EU.
30. Art. 2 of Directive 2014/65/EU refers to the members of the ESCB and other national bodies performing similar functions in the European Union, other public bodies charged with or intervening in the management of the public debt in the Union and international financial institutions established by two or more Member States that have the purpose of mobilizing funding and providing financial assistance to the benefit of their members that are experiencing or threatened by severe financing problems. This is the result of the analysis made by the European Commission in order to modify the Regulation (EU) no. 648/2012.
31. See Lemma (2013) "The Derivatives of Italy," *Law and Economics Yearly Review*, vol. 2, part. 2, p. 480 ff.
32. See *Main Results of the Council*, Luxembourg, June 9, 2009.
33. See Sinha (2013) *Regulation of Shadow Banking – Issues and Challenge*, February 1, available at http://www.rbi.org.in.
34. See Sinha (2013) *Regulation of Shadow Banking – Issues and Challenge*, February 1, available at http://www.rbi.org.in, where it is published that "the more comprehensive definition, as adopted by the Financial Stability Board (FSB), i.e., *credit intermediation involving entities and activities (fully or partially) outside the regular banking system* has been globally accepted."

    Moreover, they highlight that "globally, shadow banking entities could be covered under the broad heads of (i) Money Market Funds, (ii) Credit investment Funds, Hedge Funds, and so on, (iii) Finance Companies accepting deposits or deposit like funding, (iv) Securities brokers dependent on wholesale funding, (v) Credit insurers, financial guarantee providers and (vi) securitization vehicles."
35. See "Shadow Banking in China, Battling the Darkness," *The Economist*, May 10, 2014.
36. See "Removal of Deposit Rates' Cap Expected this Year," Chinadaily.com., March 13, 2015.
37. See Mitchell (2014) "China's Shadow Banking Loans Leap", FT.com, January 15, where it is clarified that "funding from trust companies and other

entities in the shadow sector rose to its highest level on record and accounted for 30 per cent of the Rmb17.3tn ($2.9tn) in total credit issued last year, the People's Bank of China said, up from a 23 per cent share of aggregate financing in 2012."

38. See Pozsar, Adrian, Ashcraft, and Boesky (2013) *Federal Reserve Bank of New York Staff Reports – Shadow Banking*, Federal Reserve Bank of New York Economic Policy Review, December 2013, p. 1.
39. See Pozsar, Adrian, Ashcraft, and Boesky (2013) *Federal Reserve Bank of New York Staff Reports – Shadow Banking*, cit., p. 1, where it is affirmed that "shadow banks conduct credit, maturity and liquidity transformation similar to traditional banks. However, what distinguishes shadow banks from traditional banks is their lack of access to public sources of liquidity such as the Federal Reserve's discount window, or public sources of insurance such as Federal Deposit Insurance. The emergency liquidity facilities launched by the Federal Reserve and other government agencies' guarantee schemes created during the financial crisis were direct responses to the liquidity and capital shortfalls of shadow banks. These facilities effectively provided a backstop to credit intermediation by the shadow banking system and to traditional banks for their exposure to shadow banks."
40. See Pozsar, Adrian, Ashcraft, and Boesky (2013) *Federal Reserve Bank of New York Staff Reports – Shadow Banking*, cit., p. 3 ff. for an overview of the shadow banking system. In this paper, they refer to "Pozsar (2008) and Adrian and Shin (2009). Pozsar (2008) catalogues different types of shadow banks and describes the asset and funding flows within the shadow banking system." The authors also highlight that "Adrian and Shin (2009) focus on the role of security brokers and dealers in the shadow banking system, and discuss implications for financial regulation," given that "the term 'shadow banking' was coined by McCulley (2007). Gertler and Boyd (1993) and Corrigan (2000) are early discussions of the role of commercial banks and the market based financial system in financial intermediation."
41. See Kim (2014) "Money is Rights in Rem: A Note on the Nature of Money," cit., for an analysis of the monetary impact of the shadow banking.
42. This assesses the absence of prudential rules able to mitigate the effects of an incorrect estimate of the correlation of prices (by the credit rating agencies, risk managers, and especially by investors).

    The FED observes that "specifically, they did not account for the fact that the prices of highly rated structured securities become much more correlated in extreme environments than in normal times"; See Pozsar, Adrian, Ashcraft, and Boesky (2013) *Federal Reserve Bank of New York Staff Reports – Shadow Banking*, cit., pp. 2–3.
43. See Pozsar, Adrian, Ashcraft, and Boesky (2013) *Federal Reserve Bank of New York Staff Reports – Shadow Banking*, cit., p. 13.
44. In other words, the speed of high-frequency algorithmic trading influences the market results. The European legislators specify that we are in front of a way of negotiation "where a trading system analyses data or signals from the market at high speed and then sends or updates large numbers of orders within a very short time period in response to that analysis."

    I refer to high-frequency algorithmic trading that does not require human intervention for each individual trade or order and so, it is characterized, among others, by high message intra-day rates which constitute orders, quotes or cancellations. This is why the director highlights that "in determining what constitutes high message intra-day rates, the identity of the client ultimately behind the activity, the length of the observation period, the comparison with the overall market activity during that period and the relative concentration or fragmentation of activity should be taken into account"; see *Recital* no. 61, Directive 2014/65/EU.

45. Considering what is written in the text, it can be said that the intermediation activities placed in the shadow banking system can supply the demand for credit; see European Commission, *Communication from the Commission to the Council and the European Parliament, Shadow Banking – Addressing New Sources of Risk in the Financial Sector*, cit., p. 4.

46. See Davidoff and Solomon (2007) "Black Market Capital," *Wayne State University Law School Research Paper*, cit., in which certain ramifications of black market capital have been identified and examined. The authors focus, in particular, on the current hedge fund and private equity US regulation, one likely harmful to US capital markets. This is why, from his perspective, the author finds external costs inherent in the current regulatory scheme, which the SEC have not recognized.

   It is useful to highlight that, before the crisis, there was the tendency to suggest that the supervisors (and, in particular, the SEC) should consequently undertake a thorough cost–benefit analysis of its hedge fund and private equity regulation, and the conclusion that the benefits of a regulatory scheme permitting the public offer of hedge funds and private equity funds not only exceeds its costs but is superior to current regulation.

   This is why the "black market capital" was interpreted as an example of the "unintended effects of regulating under the precautionary principle" in an era of market proliferation.

47. The issue arising from the "social revolution" developed between 1945 and 1990, which was financed by new taxes and then grew under the unfair practice of tax evasion, is not new; see Hobsbawn (1997) *Age of Extremes. The Short Twentieth Century* (Milano, Italian edition.), p. 339 ff.

48. See Fitoussi and Laurent (2008) *La nuova ecologia politica* (Milano), p. 65 ff.

49. I refer, in particular, to *Recital* no. 11, Directive 2014/65/UE, where it is specified that "a range of fraudulent practices have occurred in spot secondary markets in emission allowances (EUA) which could undermine trust in the emissions trading scheme, set up by Directive 2003/87/EC... and measures are being taken to strengthen the system of EUA registries and conditions for opening an account to trade EUAs."

   It is also clear that these measures aim "to reinforce the integrity and safeguard the efficient functioning of those markets, including comprehensive supervision of trading activity."

50. This is obviously an unfair application of the shadow processes, which is only aimed at showing a different classification of a portion of the assets (the initial loan), and, as such, not in line with the provisions of the legislation.

51. See EBA (2014) *Why Do We Need a Single Rulebook?*, available at http://www.eba.europa.eu, data accessed November 4.

2 The Shadow Banking System as an Alternative Source of Liquidity

1. See Brogi (2014) "Shadow Banking, Banking Union and Capital Markets Union," *Law and Economics Yearly Review*, 2014, p. 383 ff.
2. See Claessens, Pozsar, Ratnovski, and Singh (2012) *Shadow Banking: Economics and Policy*, edited by the IMF, December 4, p. 8.

    From a legal perspective, the result of the analysis of Alpa (2010), *Markets and Comparative Law* (London), p. 119 ff. must be taken into account, which highlights the role of regulations and moral suasion in commercial contracts and services.
3. See Anand and Sinha (2012) *Regulation of Shadow Banking – Issues and Challenge*, cit. In line with this interpretation, there is Tucker (2012) *Shadow Banking: Thoughts for a Possible Policy Agenda*, cit., which explains that "non-bank intermediation of credit is not a bad thing in itself. Indeed, it can be a very good thing, helping to make financial services more efficient and effective and the system as a whole more resilient. We must remember that as we make policy," (p. 2).
4. This is a condition—the cost reduction—that appears to be able to increase the competition between the market-based financing and the traditional credit institutions even when the offering of money has the same Value-at-Risk (that is compatible with the paradigms of capital adequacy). But, this is not only the main parameter, given the importance of the qualitative elements in the regulation of financial markets, see Capriglione (2010) *Misure anticrisi tra regole di mercato e sviluppo sostenibile*, cit., p. 83 ff. on the role of ethics in finance.
5. See Manzocchi and Padoan (2005) "The Role of Financial Markets in Economic Performance," in Boyd (ed.), *European—American Trade and Financial Alliances* (Cheltenham) p. 1 ff.
6. The positive externalities of the financial activities and credit intermediation must also be taken into account; on this point, the analysis made by Castiello (1989), "Liberalizzazione dell'attività bancaria ed evoluzione dei controlli pubblici," *Bancaria*, p. 9 ff. should be highlighted; Salanitro (1988) "Tecniche giuridiche di individuazione e regolamentazione dell'attività bancaria e finanziaria," *Banca borsa e titoli di credito*, p. 325 ff.; Fazio (1987) "Controllo dell'attività bancaria e dell' intermediazione finanziaria," *Politica del diritto*, p. 445 ff.; Castiello D'Antonio (1987) "Evoluzione dell'oggetto e qualificazione dell'attività bancaria," *Rivista del diritto commerciale*, p. 155 ff.

    Please refer also to Stagno D'Alcontres (1987) "Attività bancaria come attività di impresa e servizio pubblico in senso oggettivo," *Rivista del diritto commerciale*, II, p. 233 ff. and Scuderi (1989) "Pubblico servizio e attività bancaria," *Giurisprudenza di merito*, p. 371 ff., which develop certain ideas on the juridical definition of banking in relation with the first innovations made by the European integration and the banking directives on the Italian legal framework. In this context, there are also the results of the analysis made by Caianiello (1985) "Attività bancaria e nozione di pubblico servizio," *Il Foro italiano*, c. 130 ff.

    With regard to the juridical nature of banking after the harmonization of European legal framework, see Capriglione (2012) "Commento sub art.

10 tub," in Capriglione (ed.), *Commentario al testo unico delle leggi in materia bancaria e creditizia* (Padova 2012), I, p. 114 ff.

7. In analyzing these aspects, however, it is clear that the essence of banking is the capacity of the inter-mediators to link persons who are in surplus with others that are able to repay and reimburse the capital temporary transfer; see Lemma (2013) "Etica e professionalità bancaria," in Sabbatelli (ed.), *Etica e professionalità bancaria* (Padova), p. 129 ff.

8. See Claessens, Pozsar, Ratnovski, and Singh (2012) *Shadow Banking: Economics and Policy*, IMF Research Department, December 4, p. 10, where it is explained that "today a large part of demand for savings instruments comes from corporations and the asset management complex. Global corporate short-term savings grew from less than $50 billion in 1990 to more than $750 billion in 2007 and over $1.2 trillion by the end of 2010."

This process of "reverse maturity transformation" led the asset management complex cash holdings to rise from $100 billion in 1990 to over $2.5 trillion in 2007 and $2 trillion at end-2010, as seen in Thomas and Lemer (2011) "U.S. Groups Hit as Tax Keeps Cash Overseas," *Financial Times*, July 27.

9. In addition to the above-cited Claessens, Pozsar, Ratnovski, and Singh (2012) *Shadow Banking: Economics and Policy*, cit., there are the analyses of the FED, which observed that "large Cash Pools have created the demand for safe money in large denominations[hence]inside money, represented by repos, securities lending programs, MMMFs, and bank deposits have all expanded to meet the supply of large cash pools, but are all risky and led to costly runs"; see McAndrews (2012) "Inside and Outside Liquidity Provision," *ECB Workshop on Excess Liquidity and Money Market Functioning*, November 19/20, 2012.

10. See Claessens, Pozsar, Ratnovski, and Singh (2012) *Shadow Banking: Economics and Policy*, cit., p. 25.

11. On this point, see Monti (2001) "La dimensione internazionale della politica di concorrenza europea," *Mercatoconcorrenzaregole*, f. 3, p. 423 where the author suggests that the European institution must observe the global market (and not only the European one).

12. See Kim (2014) "Money is Rights in Rem: A Note on the Nature of Money," cit., where it is explained that shadow banking creates money by mirroring credit in the image of rights in rem.

13. It is not a potshot the assessment of the role of *backup financial institutions* to the shadow banking given that "other non-banking finance entities—such as mutual funds, insurance companies, etc.—provide alternatives to bank deposits and constitute alternative funding for the real economy, which is particularly useful when traditional banking or market channels become temporarily impaired"; see Anand and Sinha (2012) "Regulation of Shadow Banking – Issues and Challenge," cit.

14. See Tyson and Shabani (2013) "Sizing the European Shadow Banking System: A New Methodology," cit., where the authors present the results of a new methodology, used to estimate the UK shadow banking system (including European business managed from the United Kingdom). It is useful to take into account that they estimate the size at £548 billion, which, if combined

with hedge fund assets of £360 billion, gives total shadow banking assets of over £900 billion.
15. This is a process where the *asset backed securities* allow a specific structuring of the financing (by applying the techniques of *pooling* and *warehousing* in order to set the conditions required for the issuing to *collateralized debt obligations*).
16. Hence, the need to verify if the applicable rules are able to reduce the information asymmetries and the other market failures that penalize the circulation of money
17. See Stiglitz (2001) "Information and the Change in the Paradigm in Economics," *Prize Lecture*, December 8, p. 474, where it is said that "there are asymmetries of information between those governing and those governed, and just as markets strives to overcome asymmetries of information, we need to look for ways by which the scope for asymmetries of information in political processes can be limited and their consequences mitigated."
18. See Di Cagno and Spallone (2010) "An Experimental Investigation on Optimal Bankruptcy Laws," *European Journal of Law and Economics*, vol. 30, p. 205 ff.
19. See Keys, Mukherjee, Seru, and Vig (2010) "Did Securitization Lead to Lax Screening? Evidence from Subprime Loans," *The Quarterly Journal of Economics*, no. 125, p 307 ff., where the authors try to solve the question of whether the securitization process reduced the incentives of financial intermediaries to carefully screen borrowers.
20. See Bhattacharya, Chabakauri, and Nyborg (2012) "Securitized Banking, Asymmetric Information and Financial Crisis: Regulating Systemic Risk Away," available at http://www.lse.ac.uk, accessed November 4, 2014.
21. See, on this point, Stefanelli (2009) "Problematiche in ordine alla efficacia della regolazione pubblica in materia di informazione finanziaria," *Il diritto dell'economia*, p. 297 ff.
22. See Murphy (2013) *Shadow Banking: Background and Policy*, edited by Federation of American Scientist, December 31, p. 12.
23. It is useful to recall the result of an analysis on the "financial behavior"; see Marchisio and Morera (2012) "Finanza, mercati, clienti e regole, ma soprattutto persone', *Analisi Giuridica dell'Economia*, p. 19 ff.
24. See the well-known Akerlof (1970) "The Market for 'Lemons': Quality Uncertainty and the Market Mechanism," *Quarterly Journal of Economics*, p. 488 ff.
25. Thus, the analysis of the informative–cognitive relations must be conducted on an individual base, because of the various ways able to conclude a direct relationship between the informer and the informed; however, the opportunistic behavior of the first will damage only the latter in the first instance, and then it will corrupt the confidence of the investors (and, in this way, damaging the whole capital market).
26. See Mirone (2014) "Sistema e sottosistemi dellanuova disciplina della trasparenza bancaria," *Banca borsa e titoli di credito*, f. 4, p. 377 ff.
27. See Bainbridge (2005) "Shareholder Activism and Institutional Investors," *UCLA School of Law, Law-Econ Research Paper* no. 05-20, where the author outlines that institutional investor activism, by itself, will not heal the pathologies of the corporate governance.

28. See Gorton and Metrick (2010) "Regulating the Shadow Banking System," *Brookings Papers on Economic Activity*, p. 216, where the authors "propose the use of insurance for MMMFs, combined with strict guidelines on collateral for both securitization and repos, with regulatory control established by chartering new forms of narrow banks for MMMFs and securitization, and using the bankruptcy safe harbor to incentivize compliance on repos."
29. This appears in line with the Diamond and Dybvig model of bank runs, see Diamond and Dybvig (1983) "Bankruns, Depositinsurance, and Liquidity," *Journal of Political Economy*, no. 91, p. 401 ff.
30. I refer to the definition of "shadow banking system," which highlights that it is "the system of credit intermediation that involves entities and activities outside the regular banking system"; See European Commission, *Green Paper on Shadow Banking*, cit., p. 3.
31. This explains also the success of the consultation process: "the Commission received in total 140 contributions, of which 24 from Public Authorities; 47 from registered organisations; and, 64 from individual organisations. Five organisations asked for their submissions to remain confidential"; See European Commission (2012) *Summary of Responses Received to the Commission's Green Paper on Shadow Banking*, Brussels, December, p. 3.
32. This seems in line with the essence of banking, see Lemma (2013) "Etica e professionalità bancaria," cit., p. 129 f.,
33. See Keynes (1953) "The General Theory of Employment, Interest, and Money" (San Diego, edition 1964), p. 158.
34. Only in this way we can avoid convergences between these two systems that, concretely, will only align the way of acting (with obvious anti-competitive effects).
35. See Pozsar, Adrian, Ashcraft, and Boesky (2012) *Federal Reserve Bank of New York Staff Reports – Shadow Banking*, cit., p. 13 ff.
36. See Pozsar, Adrian, Ashcraft, and Boesky (2012) *Federal Reserve Bank of New York Staff Reports – Shadow Banking*, cit., whereby it is clarified also that "over the past thirty years or so, these four techniques have became widely adopted by banks and non-banks in their credit intermediation and funding practices."
37. See Pozsar, Adrian, Ashcraft, and Boesky (2012) *Federal Reserve Bank of New York Staff Reports – Shadow Banking*, cit., where the authors add that "the vertical and horizontal slicing of credit intermediation is conducted through the application of a range of off-balance sheet securitization and asset management techniques (see Exhibit 5), which enable FHC-affiliated banks to conduct lending with less capital than if they had retained loans on their balance sheets."

Moreover, the same highlights that "thus, whereas a traditional bank would conduct the origination, funding and risk management of loans on one balance sheet (its own), an FHC (1) originates loans in its bank subsidiary (2) warehouses and accumulates loans in an off-balance sheet conduit that is managed by its broker-dealer subsidiary, is funded through wholesale funding markets, and is liquidity-enhanced by the bank (3) securitizes loans via its broker-dealer subsidiary by transferring them from the conduit into a bankruptcy-remote SPV, and (4) funds the safest tranches of structured credit assets in an off-balance sheet ABS intermediary (a structured

investment vehicle (SIV), for example) that is managed from the asset management subsidiary of the holding company, is funded through wholesale funding markets and is backstopped by the bank."
38. See Carey, Post, and Sharpe (1998) "Does Corporate Lending by Banks and Finance Companies Differ? Evidence on Specialization in Private Debt Contracting," *Journal of Finance*, vol. 53, Issue 3, where the authors try to establish empirically the existence of specialization in private-market corporate lending, adding a new dimension to the public versus private debt distinctions now common in the literature. There is an interesting comparison of a large sample of corporate loans made by banks and finance companies. It shows the two types of intermediary, which are equally likely to finance information-problematic firms. In this paper, evidence supports both regulatory and reputation-based explanations for this specialization.
39. See FSB (2014), *Global Shadow Banking Monitoring Report 2014*, cit. p. 5.
40. See *Statement by European Commissioner for Competition Margre the Vestager on Tax State Aid Investigations*, Brussels, November 6, 2014, where the DG investigates on tax rulings, given that "if in a tax ruling, the tax authorities of a Member State accept that a tax base of a specific company is calculated in a favourable way which does not correspond to market conditions, it may give to the company a more favourable treatment than what other companies would normally get under the country's tax rules, and this could constitute State aid."
41. We must consider the classical approach made by Capriglione (1983) *L'impresa bancaria tra controllo e autonomia* (Milano), p. 290 ff. on the weight of limitations to banking.
42. It must be highlighted that we are in a situation different from the one examined by Schneider and Williams (2013) *The Shadow Economy*, cit., where there is a measurement of the shadow economy that requires estimation of economic activity that is deliberately hidden from official transactions, indicating that the shadow economy constitutes approximately 10 per cent of GDP in the UK, about 14 per cent in Nordic countries, and about 20–30 per cent in many southern European countries.
43. See Pozsar, Adrian, Ashcraft, and Boesky (2012) *Federal Reserve Bank of New York Staff Reports – Shadow Banking*, cit., p. 11, where it is stated that "through this intermediation process, the shadow banking system transforms risky, long-term loans (subprime mortgages, for example) into seemingly credit-risk free, short-term, money-like instruments, stable net asset value (NAV) shares that are issued by 2(a)-7 money market mutual funds which require daily liquidity."
44. It is important to consider that the third millennium started with a wave of mergers and acquisitions in the banking industry, followed by a restatement of the EU regulation (by Directive 2007/44/EC); see on this point Sciarrone and Alibrandi (2008) "Nuove regole europee in materia di acquisizioni e concentrazioni nel settore finanziario," *Banca borsa e titoli di credito*, I,p. 246 ff. It goes without saying that, nowadays, we must consider the prospective of returning to a smaller average size of banks, given the issues raised by the G-SIBs.

45. See Masciantonio (2013) "Identifying and Tracking Global, EU and Eurozone Systemically Important Banks with Public Data," *Bank of Italy Occasional Paper* no. 204.

    With regard to the G-SIIs, BCBS published in November 2014 a list of financial institutions with the corresponding "bucket" of higher loss absorbency capital. Other indicators from global systemically important institutions were issued by EBA in late September. The authority also affirms that their "identification, which leads to a higher capital requirement, falls within the responsibility of national competent authorities and will take place in January 2015 for the first time. It will follow global denominators disclosure and G-SIB exercise results, expected to be published by the BCBS and the FSB, in November each year."

46. See Masera (2012) "CRAs: Problems and Perspectives," *Analisi Giuridica dell'Economia*, Vol. 2, p. 425 ff.

47. The abovementioned consideration recalls the analysis of Amato (2013) "Il costituzionalismo oltre i confini dello Stato—Constitutionalism beyond the State," *Rivista trimestrale di diritto pubblico*, f. 1p. 1 ff., where the author deals with the emergence of powers in the global arena.

48. See Di Gaspare (2011) *Teoria e critica della globalizzazione* (Padova), p. 253 ff. on the role of globalization in the evolution of financial markets. See also Masera, "Taking the Moral Hazard out of Banking: the Next Fundamental Step in Financial Reform," *PSL Quarterly Review*, 2011, 64(257), p. 105 ff.

49. See Padoan (2013) "Criminalità organizzata, attività illegali nel sistema finanziario, paradisi fiscali e sviluppo economico," *Seminario della Scuola di Perfezionamento per le Forze di Polizia*, Roma, 8 maggio 2013; see also OECD (2013) "OECD Integrity Review of Italy. Reinforcing Public Sector Integrity, Restoring Trust for Sustainable Growth," *OECD Public Governance Reviews*, 2013.

    It is imperative to highlight the example given by Italy, where law no. 190 of 2012 provides new tools to fight against corruption within the government, followed by the Law Decree no. 101 of 2013, converted with amendments by Law no. 125 of2013, which increases the number of members and the relevance of the National Anti-Corruption Authority for evaluation and transparency of public administrations.

    See also Amorosino (2014) "Il piano nazionale anticorruzione come atto di indirizzo e coordinamento amministrativo," *Nuove Autonomie*, f. 1, p. 21 ff., where the author takes in to consideration the nature and role of the "Italian anti-corruption plan," and concludes that it is an act of administrative directive and coordination—defined as an "act of directive"—highlighting its deficiencies and limits of effectiveness.

50. Another order of issues refers to the financial innovation, which—usually through the use of derivatives—does not reduce the costs of intermediation; see Masera (2013) "Corporate governance, compliance e risk management nelle grandi banche internazionali: attività illegali e illecite, multe, indennizzi e processi penali," *Rivista trimestrale di diritto dell'economia*, Vol. 2, p. 84, where the author recalls the attention on the fact that certain international banks have operated on the CDS market in order to avoid the rules and the standard of the Basel Agreements (with short credit and, then, moving forward the risk buckets in order to reduce the capital requirements). It must

be noted that key conclusions of the cited article are: (i) official quantitative analysis of these phenomena would be highly desirable; and (ii) banks' supervisory authorities should be empowered to enact prompt incisive revision processes, with a view to ensuring the necessary standards of integrity and compliance in case of evident failures in the corporate governance of large international banks.

See also, Slovik (2013) "Systemically Important Banks and Capital Regulation Challenges," edited by OECD (Paris); Financial Crisis Inquiry Commission (FCIC), "The Financial Crisis Inquiry Report: Final Report of the National Commission on the Causes of the Financial and Economic Crisis in the United States," chaired by Phil Angelides, January 2011.

51. Even in the emerging economies the fight against illegal trade is strong, where there is a specific contrast to the distortions coming from the cash flows injected by illicit networks. It is worth noting that "From Wall Street to other financial centers across the globe, illicit networks are infiltrating and corrupting licit markets, reducing productivity, and dis-incentivizing investments in research and development—not to mention, jeopardizing public health, emaciating communities' human capital, and eroding the security of our institutions and destabilizing fragile governments," see Luna (2012) "The Destructive Impact of Illicit Trade and the Illegal Economy on Economic Growth, Sustainable Development, and Global Security," *Remarks of the Director for Anticrime Programs, Bureau of International Narcotics and Law Enforcement Affairs, The OECD High-Level Risk Forum*, OECD Conference Centre Paris, France, October 26, 2012.

52. I am referring to problems similar to the ones that characterized the "London Whale". As reported by Hurtado (2014) "The London Whale," *Bloomberg Quick Tale*, updated October 21, 2014. The Whale case received attention when the US Federal Reserve's Inspector General said that "regulators had botched oversight of the JPMorgan unit where the losses took place." More in particular, the examiners in the New York Fed had spotted risks in the unit's trading as early as 2008 but never followed up. Furthermore, the cited article precise that "there was poor coordination with other regulatory agencies."

See, also, Bart and McCarthy (2012) "Trading Losses: A Little Perspective on a Large Problem," *Milken Institute*, October, 2012, and the report *ad hoc JPMorgan Chase Whale Trades: A Case History of Derivatives Risks and Abuses*, March 15, 2013.

53. See *CEBS Guidelines On Prudential Filters for Regulatory Capital*, December 2004. I refer, in particular, to the role that—in Europe—has been recognized as the "prudential filters," which are the rules able to limit the negative effects of certain accounting standards and principles; see also Lemma (2006) "L'applicazione del Fair Value alle banche: problematiche giuridiche e soluzioni," *Banca borsa e titoli di credito*, I, p. 723 ff.

54. In this direction is oriented the work of Masera (2007) "L'impresa e la creazione di valore," in Capriglione (ed.), *Finanza, Impresa e Nuovo Umanesimo* (Bari), p. 67 ff.

55. See Capriglione (2007) "Introduzione" and Antonucci (2007) "La responsabilità sociale e l'impresa bancaria," in *Finanza, Impresa e Nuovo Umanesimo*, cit., p. 14; see also Baggio (2005), *Etica ed economia* (Roma), p. 19.

56. See Levy (2006) *The State after Statism: New State Activities in the Age of Liberalization* (Cambridge), p. 469 ff.

3 Shadow Banking Entities

1. See Gorton and Souleles (2005) "Special Purpose Vehicles and Securitization," *FRB Philadelphia Working Paper*, no. 05-21, where the authors analyze securitization and special purpose vehicles (SPVs), which are more common in corporate finance praxis.
2. See Fabozzi and Kothari (2007) "Securitization: The Tool of Financial Transformation," *Yale ICF Working Paper*, no. 07-07, where the authors find that, in a broadest sense, the term "securitization" implies a process by which a financial relationship is converted into a transaction.
3. See Lucas, Goodman, and Fabozzi (2007) "Collateralized Debt Obligations and Credit Risk Transfer," *Yale ICF Working Paper*, no. 07-06, where the authors emphasize the role played by CDOs in the application of the securitization technology and in the credit risk transformation schemes.
4. See Coval, Jurek, and Stafford (2008) "The Economics of Structured Finance," *Harvard Business School Finance Working Paper*, no. 09-060, where the authors examine how the process of securitization allowed a big amount of risky assets to be transformed into securities that were widely considered to be safe. They argue that at the core of the financial market crisis has been the discovery that these securities are actually far riskier than originally advertised.
5. It goes without saying that the link mentioned in the text suggests the inclusio of SPVs within the consolidation range of the relevant banking group (in a capital adequacy perspective), for a pre-crisis analysis of the securitization legal framework that explains this problem, see Troiano (2003) *Le operazioni di cartolarizzazione* (Padova), where the author takes into consideration both the EU regulation and the Italian legislation.
6. It shall be taken into account thatoften the vehicles act in accordance with executive directions or legal frameworks processed by the banks' managers on the basis of their technical and financial knowledge, and the assumption related to the demand for securities coming from its clients.
7. Often, the lack of internal control systems has prevented the assumption of appropriate decisions in front of the deterioration of the assets owned. And this went along with consequent systemic degenerations.
8. It shall be useful to take into account the Italian praxis of supervision on the consolidation of the balance sheets, see Bank of Italy, "Disposizioni di vigilanza per le banche," Circular no. 285 of December 17, 2013, First Part.I.2.7.
9. See Becht, Bolton, and Röell (2002) "Corporate Governance and Control," *ECGI—Finance Working* Paper, no. 02/2002, where the authors identify a specific corporate governance dilemma whether the intervention of a large shareholder needs to be regulated in order to guarantee a better protection for small investors, taking into account that such a rule increases managerial discretion and scope for abuse.

10. See Guida and Masera (eds) (2014) *Does One Size Fit All?*, *passim*, where the authors state that "an important lesson of the global financial crisis was that the Basel II microprudential capital requirements did not ensure financial stability." Another significant lesson was that they had even destabilizing effects by increasing procyclical lending and regulatory arbitrage.
11. See Tosun and Senbet (2014) "Internal Control and Maturity of Debt," *WBS Finance Group Research Paper*, where they investigate the various effects of internal board monitoring on firms' debt maturity structure.
12. See, in particular, Masera (2014) "CRR/CRD IV: The Trees and the Forest," *SSRN Working Paper* no. 2418215.
13. See, in particular, the provisions of Regulation EU no. 575 of 2013, part three, Tit. II.5 and part five.
14. See Babis (2014) "Single Rulebook for Prudential Regulation of Banks: Mission Accomplished?", *European Business Law Review* (Forthcoming), *University of Cambridge Faculty of Law Research* Paper, no. 37/2014, where the author focuses on the prudential regulation aspects of the rulebook, examining whether the CRR/CRD 4 framework has created a truly single rulebook by identifying possible threats to uniformity.
15. See Keynes (1953) *The General Theory of Employement, Interest, and Money*, cit., pp. 338–339.
16. See Adrian andLiang (2014) "Monetary Policy, Financial Conditions, and Financial Stability," cit., where the authors show the consequences that arises if monetary policy does not target financial stability considerations and the financial vulnerabilities given from persistent accommodative monetary policy.
17. See art. 14, EU Regulation no. 1024/2013.
18. Therefore, the set named as shadow banks must be referred, for its content, to a phenomenon with a plural nature and a composite outlook, which can complement commercial banks according to the "genus-differentia" scheme; see Webber (1991) *L'etica protestante e lo spirito del capitalismo*, (Milano edition, 1994), p. 71.
19. See Pozsar, Adrian, Ashcraft, and Boesky (2012) *Federal Reserve Bank of New York Staff Reports – Shadow Banking*, cit., p. 1, where it is highlighted that the specialized shadow bank intermediaries are bound together along an intermediation chain. They, in particular, reveal the presence of a "network of shadow banks," operating in this intermediation chain as the shadow banking system. Therefore, the authors believes that shadow banking is a somewhat pejorative name for such a large and important part of the financial system.
20. See Schwarcz (2013) "Regulating Shadows: Financial Regulation and Responsibility Failure," *Washington and Lee Law Review* (forthcoming), where the author highlights the modern financial architecture, where financial services are increasingly provided outside of the traditional banking system and without the need for bank intermediation between capital markets and the users of funds.
21. It can be useful to consider the recent statements on the distribution of complex financial products published by ESMA; see *Opinions* of February 7, 2014 (MiFID practices for firms selling complex products) and March 27, 2014 (Structured Retail Products—Good practices for product governance

arrangements). Please note that Italian supervisory authority CONSOB published a Communication (N. 0097996/14 of December 22, 2014) in order to identify a list of "complex financial products" (including asset backed securities, convertible, structured, and credit linked products) and advises intermediaries not to offer such products to retail investors.
22. See Meeks, Nelson, and Alessandri (2013) "Shadow Banks and Macroeconomic Instability," *Bank of Italy Working Papers*, no. 939, p. 19 ff. and see also Adrian and Ashcraft (2012) "Shadow Banking Regulation," *FRB of New York Staff Report*, no. 559 where the authors highlight how shadow banks conduct credit intermediation without direct or explicit access to public funding or other credit guarantees.
23. See, once more, Meeks, Nelson, and Alessandri (2013) *Shadow Banks and Macroeconomic Instability*, cit., p. 11.
24. In this sense, moreover, are oriented international institutions that recognize the existence of a plurality of subjects, autonomous in the adoption of appropriate corporate governance systems and in the definition of their capital structures; see Bakk-Simon, Borgioli, Giron, Hempell, Maddaloni, Recine, and Rosati "Shadow bank in the Euro Area: an Overview," edited by ECB, *Occasional Paper*, no. 133/April 2012.
25. See McCulley (2014) "Make Shadow Banks Safe and Private Money Sound," FT.com, June 16.
26. See Goodhart, Kashyap, Tsomocos, and Vardoulakis (2012) "Financial Regulation in General Equilibrium," *Chicago Booth Research* Paper, no. 12-11, where the authors analyze how different types of financial regulation combat the financial crisis. They highlight in the Abstract that "the proposed framework can assess five different policy options that officials have advocated for combating defaults, credit crunches and fire sales, namely: limits on loan to value ratios, capital requirements for banks, liquidity coverage ratios for banks, dynamic loan loss provisioning for banks, and margin requirements on repurchase agreements used by shadow banks."
27. See Zingales (2009) "The Future of Securities Regulation," *Chicago Booth School of Business Research Paper* no. 08-27, where the author analyzes investors' confidence after a crisis, comparing the Great Depression of the 1929 and the recent financial crisis of 2007.
28. See, on this topic, Pennisi (2009) "La responsabilità della banca nell'esercizio del controllo in forza di covenants finanziari," *Rivista di diritto societario*, f. 3, pt. 3, p. 627 ff., for an analysis of the duites provided by the Italian legislation.
29. In order to better understand the benefits arising from the use of captive financial company, we must consider the possibility to eliminate independent credit institutions and services providers. So the holding of the group can set up the loans according to its business parameters and not (only) to the financial market trends; see on this point Bodnaruk, Simonov, and O'Brien (2012) "Captive Finance and Firm's Competitiveness," *SSRN Working Paper*, no. 2021503.
30. See Moodys' *The Fundamentals of Asset-Backed Commercial Paper*, February 3, 2003, available at http://www.imf.org.
31. I refer to the analysis of Adrian (2011) "Dodd-Frank One Year On: Implications for Shadow Banking," *Federal Reserve Bank of New York Staff Report*,

no. 533, p. 3 where the question of the consolidation of ABCP conduits is analyzed.
32. See Commission Decision of June 4, 2008 on State Aid C 9/08 (ex NN 8/08, CP 244/07) where it is specified that a SIV is "a corporate body... have been used because they can remain off balance sheet and not be consolidated by banks. This allows banks tofund lending at cheaper rates than those they would provide themselves (due notably to the obligations of regulatory liquidity ratios). The conduit refinances investments in asset-backed securities (ABSs) by borrowing in the short-term asset-backed commercial paper (CP) market. Potential liquidity needs of the conduits (where the commercial papers are not sol dcompletely) are bridged by credit lines provided by commercial banks."
33. See Commission Decision of June 4, 2008, para. 8, cit. where it is stated that "Ormond Quay generated a significant surplus by financing the long-term and high-yield ABS investments through short-term and low-yield commercial papers," therefore, the European Commission has adopted the following decision: "the liquidity facility and the guarantee granted to Landesbank Sachsen Girozentrale (Sachsen LB) in connection with its sale constitute state aid within the meaning of article 87(1) EC that is compatible with the common market subject to the obligations and conditions set out in article 2" (art. 1).
34. See Claessens, Pozsar, Ratnovski, and Singh (2012) "Shadow Banking: Economics and Policy," cit., p. 22.
35. See Lemma (2006) *I fondi immobiliari tra investimento e gestione* (Bari), p. 197 ff.
36. See ECB, *Guideline of the European Central Bank*, of September 20, 2011, p. 14, note 6.
37. See SEC, *Money Market Funds*, definition available at http://www.sec.gov.
38. See EC, *Money Market Funds*, Proposal of the Commission, of September 4, 2013, and European Commission, *New Rules for Money Market Funds Proposed*, Memo/13/764 of September 4, 2013.
39. See Tarullo (2013) "Shadow Banking and Systemic Risk Regulation," speech at the Americans for Financial Reform and Economic Policy Institute Conference, Washington, DC, November 22, 2013.
40. See EC, *Money Market Funds*, Proposal of the Commission, of September 4, 2013, cit.
41. See EC, *New Rules for Money Market Funds Proposed – Frequently Asked Questions*, cit.
42. See Carney (2014) "Taking Shadow Banking Out of the Shadows to Create Sustainable Market-Based Finance," *Financial Times*, June 16, 2014, where it is said that "the cycle of excessive borrowing in economic booms that cannot be sustained when liquidity dissipates in core fixed income markets."
43. See Tucker (2012) "Shadow Banking: Thoughts for a Possible Policy Agenda," speech at the European Commission High Level Conference, Brussels, April 27, 2012, p. 4, where it is highlighted that "Compared to most types of shadow banking, money funds do not borrow – in the usual sense. But by promising par, they are in effect incurring debt-like obligations. And they can be exposed to leverage. At least in the run up to the crisis, some invested in levered paper, some of it in what amounted to Russian Doll

shadow banking – a money fund buys short-term ABCP backed by CDOs, etc."
44. See Carney (2014) "Taking Shadow Banking Out of the Shadows to Create Sustainable Market-Based Finance," cit.

4 Shadow Business of Banks, Insurance Companies, and Pension Funds

1. On the competition in the financial market and the need to clarify the relationship between the first and the need for a stable equilibrium (of the second), see Rabitti (2014) "La concorrenza nel settore finanziario e i provvedimenti del Governo Monti," *Assicurazioni*, II,p. 441 ff.
2. In other words, the research for the maximum efficiency in the shadow process required to provide all the services able to build a relationship between the demand for loans and the supply of structured financial instruments; see Pozsar,Adrian, Ashcraft, and Boesky (2012) *Federal Reserve Bank of New York Staff Reports – Shadow Banking*, cit., p. 11.
3. See Greene and Broomfield (2013) "Promoting Risk Mitigation, Not Migration: A Comparative Analysis of Shadow Banking Reforms by the FSB, USA and EU," *Capital Markets Law Journal*, vol. 8, no. 1, where the authors focus on the importance of tailored solutions (made to address the specific activities which create risk), rather than apply standard rules to shadow banking entities, ignoring their own characteristics or risk profiles.
4. See Adrian and Ashcraft (2012) "Shadow Banking Regulation," *FRB of New York Staff Report* no. 559, cit., where the authors review the implications of certain shadow funding sources, including asset-backed commercial paper, triparty repurchase agreements, money market mutual funds, and securitization.
5. See Pozsar, Adrian, Ashcraft, and Boesky (2012) *Federal Reserve Bank of New York Staff Reports – Shadow Banking*, cit., p. 13 ff.
6. See Desiderio (2005) "L'attività bancaria," in Capriglione (ed.) *L'ordinamento finanziario italiano* (Padova), p. 248 ff.
7. See *Recitals* nos. 2 and 5, Directive 2013/36/EU.
8. Therefore, the analysis of the supervisors will have to identify the exposure (on the way of banks) to the negative externalities of the aforementioned interactions (among the latter and the shadow banking entities); see Schwarcz (2013) "Regulating Shadows: Financial Regulation and Responsibility Failure," *Washington and Lee Law Review* (forthcoming), cit., where the author identifies the externalities as a third marketfailure category that need to be conceptualized as a sort of "responsibility failure."
9. In a future perspective, it would be useful to take into account the uniform definitions and reporting requirements for forbearance and non-performing exposures set out by the European Banking Authority on "EBA Final Draft implementing technical standards on supervisory reporting on forbearance and non performing exposures under article 99(4) of CRR."
10. See on this topic Anelli (1998) "La responsabilità risarcitoria delle banche per illeciti commessi nell'erogazione del credito," *Diritto della banca e del mercato finanziario*, f. 2,p. 137 ff.

11. See Duffie and Zhu (2011) "Does a Central Clearing Counterparty Reduce Counterparty Risk?", *Rock Center for Corporate Governance at Stanford University Working Paper*, no. 46, where the authors demonstrate how the participation of a central clearing may lower counterparty risk for a particular class of derivatives.
12. See Siclari (2013) "Tendenze regolatorie in materia di compliance bancaria," *Rivista Trimestrale di Diritto dell'Economia*, I, p. 156 ff., where the author quotes Capriglione (2009) *Crisi a confronto (1929 e 2009). Il caso italiano*, (Padova), *passim*; Napolitano (2009) "L'intervento dello Stato nel sistema bancario e i nuovi profili pubblicistici del credito," *Giornale dir. amm.*, no. 4, p. 429 ff.

    See also Lemme (2014) "Le disposizioni di vigilanza sulla governance delle banche: riflessioni a tre anni dall'intervento," *Banca borsa e titoli di credito*, 2011, f. 6,p. 705 ff.
13. See Enriques and Zetzsche (2014 "Quack Corporate Governance, Round III? Bank Board Regulation Under the New European Capital Requirement Directive," *Oxford Legal Studies Research Paper*, no. 67/2014, where the authors focus on the provisions aimed to reshape bank boards' composition, functioning, and members' liabilities. They argue that these measures are not appropriate to improve the bank's governance effectiveness and to prevent excessive risk taking.

    See also Bebchuk and Spamann (2010) "Regulating Bankers' Pay," *Georgetown Law Journal*, vol. 98, Issue 2, pp. 247–287 and *Harvard Law and Economics Discussion Paper*, no. 641, for a useful analysis of the incentive connected to banks' executive pay, which—as known—has produced excessive risk taking.

    In this study, the fact that in credit institutions pay packages focused excessively on short-term results must be taken into account, and this produces critical distortion that has received too little attention by the supervisors; see Bank of Italy, Circ. no. 285 of December 17, 2013, First Part, Title III, Ch. 1, Sec. III for an example of the national regulation of this topic.

    Undoubtedly, there is the need for corporate governance reforms aimed at aligning the design of executive pay arrangements with the interests of the industry (i.e. the stability of the market). Moreover, according to the Bebchuk and Spamann analysis, the interests of common shareholders could be served by more risktaking than is socially desirable. Thus, the results of the above analysis shall be taken into account, which provide a normative foundation for such pay regulation.
14. See Lemma (2011) "La riforma degli intermediari finanziari non bancari nella prospettiva di Basilea III," *Rivista elettronica di diritto, economia e management*, p. 184 ff., where I compare the Italian reform of financial intermediaries (enacted by legislative decree no. 141/2010) with the comprehensive set of measures developed in 2010 by the Basel Committee (i.e. Basel III). This work presents results that are useful for understandingwhether there is a link between the Italian reform and the new international "soft law" written to strengthen the regulation, supervision, and risk management of the banking sector.

    In the context of this article, the analysis of the shadow banking system shows the benefit of this new wide-ranging approach in the supervision of various financial firms (including lenders and credit guarantee consortia). Nevertheless, it must not be forgotten that the innovations enacted by legislative decree no. 141/2010 purport the option of wider supervision—equal to banks and financial intermediaries—to demonstrate that this improves the sustainability of the markets.

15. See Barbagallo (2014) "Doveri e responsabilità degli amministratori delle banche: il punto di vista della Banca d'Italia," speech at ABI conference L'impresa bancaria: i doveri e le responsabilità degli amministratori, p. 6, on the remedial action provided at European level against the crisis.
16. See Vella (2007) "Le Autorità di vigilanza: non è solo questione di architetture," *Dir. banca merc. fin.*, 2, p. 196.
17. This is linked to the document EBA, *Guidelines on Internal Governance*, September 2011, Title III, Section 23 (1, 3) where it is stated that "an institution must have in place a well-documented new product approval policy (NPAP), approved by the management body, which addresses the development of new markets, products and services and significant changes to existing ones."
18. To this purpose are the regulations on the CEO-1 risks, which appear to be able to condition the management, see Banca d'Italia, *Nuove disposizioni di vigilanza prudenziale per le banche*, Circular no. 263 of December 27, 2006 – 15° update of July 2, 2013, pp. 6–7.
19. See Billio, Lo, Sherman, and Pelizzon (2011) "Econometric Measures of Connectedness and Systemic Risk in the Finance and Insurance Sectors," *University Ca' Foscari of Venice, Dept. of Economics Research Paper Series*, no. 21 and *MIT Sloan Research Paper* no. 4774–10, where the authors propose an econometric model of connectedness, applied to the monthly returns of hedge funds, banks, brokers/dealers, and insurance companies. They find out that those sectors have become highly interrelated over the past decade, increasing the total level of systemic risk. However, the study shows a relevant asymmetry in the degree of interconnectedness among the financial system, with banks playing a much more important role in transmitting shocks than other financial institutions.
20. See Corrias (2013) "Causa del contratto di assicurazione: tipo assicurativo o tipi assicurativi?", *Rivista di diritto civile*, f. 1, p. 41 ff., on the role of the insurance contracts in the financial market
21. See Commission Delegated Regulation (EU) of October 10, 2014 supplementing Directive 2009/138/EC of the European Parliament and of the Council on the taking up and pursuit of the business of Insurance and Reinsurance (Solvency II), Brussels, October 10, 2014, C(2014) 7230 final.
22. These rules are quite different from the options made by EC, *Explanatory Memorandum*, Brussels, 10 October 2014, C(2014) 7232 final, in adopting the Capital Requirements Regulation (CRR1) in June 2013, the co-legislators introduced a requirement for all institutions (credit institutions and investment firms) to maintain a general Liquidity Coverage Requirement (article 412(1)) and to report regularly on the composition of the liquid assets in their liquidity coverage buffer to their competent authorities (CRR Articles 415–425).
23. See, *Delegated Regulation of 10th October 2014*, cit.

24. See Pozsar, Adrian, Ashcraft, and Boesky (2012) *Federal Reserve Bank of New York Staff Reports – Shadow Banking*, cit., p. 22.
25. See *Delegated Regulation of 10th October 2014*, cit., p. 8.
26. See *Delegated Regulation of 10th October 2014*, cit., p. 8, where it explains that "the Solvency II Directive empowers the Commission to specify the elements of the system of governance including a non-exhaustive list of written policies."
27. See Regulation (EU) no. 1094/2010, Recital no. 10.
28. See Gallin (2013) *Shadow Banking and the Funding of the Nonfinancial Sector*, available at http://www.nber.org, p. 7 and p. 8. See also Boersch (2010) "Doing Good by Investing Well—Pension Funds and Socially Responsible Investment: Results of an Expert Survey," *Allianz Global Investors International Pension Paper* no. 1/2010, where the author conducts research on the future of socially responsible investment in pension fund portfolios. Hence, the explaination that pension funds are one of the main drivers of socially responsible investments just because of their long-term horizon and asset size.
29. See Gallin (2013) *Shadow Banking and the Funding of the Nonfinancial Sector*, available at http://www.nber.org, p. 10.
30. See, for a pre-crisis perspective, Impavido (2002) "On the Governance of Public Pension Fund Management," *World Bank Policy Research Working Paper*, no. 2878, where the author highlights the need for further reform to support the development of an appropriate set of governance guidelines.

5 Shadow Banking Operations

1. See Fabozzi and Kothari (2007) "Securitization: The Tool of Financial Transformation," *Yale ICF Working Paper*, no. 07-07, cit., where the authors explain how the combination of securitization techniques with credit derivatives and risk transfer devices creates innovative methods of transforming risk into commodities, allowing market participants to operate into sectors which were otherwise not open to them.
2. See Tucker (2012) *Shadow Banking – Thoughts for a Possible Policy Agenda*, cit., p. 3.
3. See Adrian and Ashcraft (2012) "Shadow Banking Regulation," *FRB of New York Staff Report*, no. 559, cit., where they confirm how shadow banks contributed to the credit boom in the early 2000s and then collapsed during the recent financial crisis of 2007–2009.
4. See European Comission, *Green Paper on Shadow Banking*, cit., p. 4, where the author refers to the FSB's analysis on the size of the phenomenon: "The FSB has roughly estimated the size of the global shadow banking system at around €46 trillion in 2010, having grown from €21 trillion in 2002. This represents 25–30% of the total financial system and half the size of bank assets. In the United States, this proportion is even more significant, with an estimated figure of between 35% and 40%."

   In this context, it is interesting to consider "the share of the assets of financial intermediaries other than banks located in Europe as a percentage of the global size of shadow banking system has strongly increased from 2005 to

2010, while the share of US located assets has decreased. On a global scale, the share of those assets held by European jurisdictions has increased from 10 to 13% for UK intermediaries, from 6 to 8% for NL intermediaries, from 4% to 5% for DE intermediaries and from 2% to 3% for ES intermediaries. FR and IT intermediaries maintained their previous shares in the global shadow banks assets of 6% and 2% respectively."

5. See, on this point, Troeger (2014) "How Special Are They? – Targeting Systemic Risk by Regulating Shadow Banking," for an analysis of the possible regulatory interventions.
6. See Pozsar, Adrian, Ashcraft, and Boesky (2012) *Federal Reserve Bank of New York Staff Reports – Shadow Banking*, cit., p. 10.
7. These operations run in relation to "loan origination" and the "loan warehousing" (funded by an ABS issuance or an ABCP issuance); see, once more, Pozsar, Adrian, Ashcraft, and Boesky (2012) *Federal Reserve Bank of New York Staff Reports – Shadow Banking*, cit., p. 10, where it is assumed that "in the shadow banking system, loans, leases, and mortgages are securitized and thus become tradable instruments. Funding is also in the form of tradable instruments, such as commercial paper and repo. Savers hold money market balances, instead of deposits with banks."
8. Undoubtedly, the rights arising from the loan will have to follow the pattern of credit transfer or be realized through the mechanisms of collective portfolio management (and, therefore, through the intervention of a "fund") or in synthetic mode (by entering in "credit default swaps").
9. See the case shown by Tucker (2012) *Shadow Banking: Thoughts for a Possible Policy Agenda*, cit., p. 3.
10. In other words, according to Tucker (2012) *Shadow Banking: Thoughts for a Possible Policy Agenda*, cit., p. 3: "That is, banks should hold more liquid assets against such exposures."
11. See Pozsar, Adrian, Ashcraft, and Boesky (2012) *Federal Reserve Bank of New York Staff Reports – Shadow Banking*, cit, p. 4, where it is stated that, "for example, a pool of illiquid whole loans might trade at a lower price than a liquid rated security secured by the same loan pool, as certification by a credible rating agency would reduce information asymmetries between borrowers and savers."
12. See Capriglione (2010) *Introduzione*, in Urbani (ed.) *L'attività delle banche* (Padova), p. 1.
13. Hence, we can rely on a peculiar form of enhancements, classified on the basis of relationships that exists with major operations "as either direct or indirect, and either explicit or implicit"; see Pozsar, Adrian, Ashcraft, and Boesky (2012) *Federal Reserve Bank of New York Staff Reports – Shadow Banking*, cit., p. 5.
14. It should indeed check which are the constraints that the legislation poses to avoid the prevailance, in practice, of loan agreements to the real economy that transfer (or, rather, amplify) the risk of insolvency of the transferor.
15. In particular, the solution provided by Section 941 of the Dodd–Frank Act links the public intervention to a definition of the above operation where "a 'securitizer' to mean '(a) an issuer of an asset-backed security or (b) a person who organizes and initiates an asset-backed securities transaction by selling or transferring assets, either directly or indirectly, including through an

affiliate, to the issuer'," while the term "an 'originator' to mean a person who (a) through extension of credit or otherwise, creates a financial asset that collateralizes an asset-backed security and (b) sells an asset directly or indirectly to a securitizer"; see Board of Governors of The Federal Reserve System, *Report to the Congress on Risk Retention*, October 2010, p. 9.

See also Skeel (2010) "The New Financial Deal: Understanding the Dodd-Frank Act and its (Unintended) Consequences," cit., where it his highlighted that, even if Dodd-Frank Act will not end bailouts, it will be useful to orient the market towards safer trends.

16. See Board of Governors of The Federal Reserve System, *Report to the Congress on Risk Retention*, cit., pp. 1–2.
17. See the Preamble of law decree no. 145 of 2013, and its regulatory content as amended by the Conversion Law no. 9 of 2014.
18. See Bank of England and ECB, *The Case for a Better Functioning Securitization Market in the European Union*, May 2014, pp. 13–14.
19. See Bank of England and ECB, *The Case for a Better Functioning Securitization Market in the European Union*, cit., p. 19.
20. See Albertazzi, Eramo, Gambacorta, and Salleo (2011) "Securitization is not that Evil After All," *Bank of Italy Working Papers*, February 2011, p. 3.
21. This, however, does not explain the validity of the choice to anchor the production of shadow financial instruments to the securitization of loans to the real economy, nor clarifies which controls are to be provided to prevent the effect of the credit transformation is carried out to the detriment of final investors; see Albertazzi, Eramo, Gambacorta, and Salleo (2011) "Securitization is not that Evil After All," cit., p. 39, where it is added that "this result is consistent with the idea that the choice of the loans to be securitized is made with the aim of overcoming asymmetric information problems. Beyond this, we provide new direct evidence that the structure of the securitization deals is also chosen with a view to mitigating the costs of asymmetric information. We find evidence consistent with the fact that banks, particularly at the early stage of the securitization market life, are strongly committed to building up a reputation that will allow them to ensure repeated access to this important source of funding."

It shall be helpful to consider also the analysis on cross-country data made by these authors, because they show that the securitization of prime mortgages is a soundly functioning market, and—according to the paper—it should not be excessively penalized.
22. See Merusi (2009) "Per un divieto di cartolarizzazione del rischio di credito," *Banca borsa e titoli di credito*, I, p. 253 ff. and *Giustizia amministrativa*, p. 315 ff., where the author highlights the choice of non-facing this kind of financial activities, given that certain authorities were "investing part of the pension fund of its managers in hedge funds."
23. Furthermore, it shall be considered that an absence of capital requirements (calculated on the basis of credit risk) should have a competitive advantage for the shadow banking entities; see Kolm (2014) "Securitization, Shadow Banking, and Bank Regulation," *SSRN Working Paper*, no. 2521390.
24. I am referring not only to the new "retention rule," but also to the prospective interventions following the Directive no. 2010/76/EC, which improves the fight against the excessive and imprudent risk taking in the banking sector, and—in order to address the potentially detrimental effect of unfair

operations—a new form of supervision for credit institutions investing in re-securitization; see Recitals nos. 1, 3, and 24 of Directive no. 2010/76/EU.
25. See Adrian and Shin (2009) "The Shadow Banking System: Implications for Financial Regulation," *FRB of New York Staff Report*, no. 382,where it was anticipated that in the new, post-crisis financial system, the role of securitization will likely be held in check by more stringent financial regulation and by the recognition that it is important to prevent excessive leverage and maturity mismatch, both of which can undermine financial stability.
26. See Gorton and Metrick (2010) "Securitized Banking and the Run on Repo," *Yale ICF Working Paper*, no. 09-14, where the authors find that changes in the "LIB–OIS" spread, a proxy for counterparty risk, were strongly correlated with changes in credit spreads and repo rates for securitized bonds. They assume that these changes implied higher uncertainty about bank solvency and lower values for repo collateral. This raises concerns about the liquidity of markets for the bonds used as collateral led to increases in repo "haircuts": the amount of collateral required for any given transaction. They conclude that, because the declining asset values and increasing haircuts, the US banking system was effectively insolvent for the first time since the Great Depression.
27. See Mazzuca and Agostino (2011) "Empirical Investigation Of Securitisation Drivers: The Case of Italian Banks," *The European Journal of Finance*, Vol. 17, Issue. 8, p. 623 ff.
28. See Hansel and Krahnen (2007) "Does Credit Securitization Reduce Bank Risk? Evidence from the European CDO Market," *Working Paper*, March 20, available at http://www.cepr.org, where the authors analyze whether the use of credit risk transfer instruments affects risk taking by large, international banks.
    Furthermore, see Nijskens and Wagner (2011) "Credit Risk Transfer Activities and Systemic Risk: How Banks became Less Risky Individually but Posed Greater Risks to the Financial System at the Same Time," *Journal of Banking & Finance*, Vol. 35, Issue 6, p. 1391 ff. See also Uhde, Farruggio, and Michalak (2012) "Wealth Effects of Credit Risk Securitization in European Banking," *Journal of Business Finance & Accounting*, Vol. 39, where—using a unique cross-sectional dataset of 381 cash and synthetic securitizations issued by 53 banks from the EU-15 plus Switzerland between 1997 and 2007—these authors provide an empirical analysis of the time-dependent negative wealth effects of credit risk securitization announcements in European banking.
29. See Battaglia and Mazzuca (2012) "La relazione tra attività di cartolarizzazione e liquidità nelle banche italiane. Alcune evidenze empiriche dalla recente crisi finanziaria—Securitization and liquidity in Italian banks. Evidence from the recent financial crisis," *Banca Impresa Società*, p. 419 ff.
30. On the juridical relevance of the gentlemen's agreements, see Di Donna (2013) *Gentlemen's Agreements. Notazioni sulla fenomenologia degli accordi*, (Napoli) p. 12 ff.
31. See Picardi (2008) "Il 'Fondo comune di crediti' nel sistema della separazione patrimoniale," *Banca borsa e titoli di credito*, I, p. 76 ff.
32. See Rucellai (2012) "Cartolarizzazione sintetica e Credit Default Swap," *Giurisprudenza commerciale*, n. 3, p. 1, p. 371 ss.
33. It is important to once more recall the dynamics of "risk retention, transparency, and standardization," whose supervision must be entrusted to

authorities able to identify and manage the risks, and thus to limit the negative impact of a default of the shadow banking operations on the capital market; see Claessens, Pozsar, Ratnovski, and Singh (2012) *Shadow Banking: Economics and Policy*, cit., p. 22.
34. See Cossu—Spada (2010), "Dalla ricchezza assente alla ricchezza inesistente—divagazioni del giurista sul mercato finanziario," *Banca borsa e titoli di credito*, I, p. 401 ff.
35. This is a setting that allows us to grasp the limits of the techniques in question and the difficulties of a lawyer to deal with atypical forms of circulation of credit rights, especially with regard to cases in which there is the transfer of risks associated with the underlying assets (in terms of collection, origin and quality of the right acquired).All in all, in the shadow banking system, there are specific difficulties in managing *macro-interests* and, in particular, those with a collective character, as happens in the investment of savings and in the raising of equity; see Cossu and Spada (2010) "Dalla ricchezza assente alla ricchezza inesistente—divagazioni del giurista sul mercato finanziario," cit., para 6.2.1.
36. See Verde (2014) "Cartolarizzazione, patrimonio separato e limiti alle garanzie per gli investitori—Securitization, Separate Assets and Limits to the Guarantees for the Investors," *Rassegna di diritto civile*, p. 485 ff.
37. Moreover, it has been understood that any investor does not receive enough data to assume that his or her choices can be based on rationale decisions.
38. In addition, the financial instruments in question, as already mentioned, have to be traded in financial markets in order to be evaluated according to the criteria of fair market value (and, therefore, to produce the pro-cyclical amplification effects that the application of this method allows arrangers to realize). This makes it necessary that the placement of these instruments follows the rules laid down by the relevant regulatory framework.
39. See Ammannati (2012) "Mercati finanziari, società di rating, autorità ed organismi di certificazione," *Rivista di Diritto Alimentare*, 2012, f. 1, p. 17 ff., where the author clarifies that experience of the financial markets shows how the economic operators are not always structured on the rational assessment of information received, and then she specifies that the rating, as expressed in an alphabetical and easily intelligible mark, has partially changed its value in the market and is aimed at spreading systemic trust based on the CRAs' reputations.
40. See Partnoy (2006) "How and Why Credit Rating Agencies are Not Like Other Gatekeepers. Financial gatekeepers: can they protect investors?," *San Diego Legal Studies Paper*, no. 07-46, where the author describes how CRAs have changed radically since 1999, becoming more profitable than other gatekeepers and facing different and more important conflicts of interest. The author also highlights how CRAs are uniquely active in structured finance and particularly on collateralized debt obligations.
41. See Troisi (2013) *Le Agenzie di rating* (Padova), p. 152; see also Giudici (2011) "L'agenzia di rating danneggia l'emittente con i propri rating eccessivamente favorevoli?," *Le Società*, f. 12, p. 1451 ff.
42. See Troisi (2014) "Rating e affidamento dell'investitore: profili di responsabilità dell'agenzia," *Rivista Trimestrale di Diritto dell'economia*, II, p. 166 ff., where the author comments the decision of the *Federal Court of Australia*,

ABN AMRO Bank NV v Bathurst Regional Council [2014] FCAFC 65 related to this topic.
43. See Lener and Rescigno (2012) "Agenzie di "rating" e conflitti di interesse: sintomi e cure i," *Analisi Giuridica dell'Economia*, f. 2, p. 353 ff. for a deeper analysis of the conflict of interests.
44. See Troisi (2013) *Le Agenzie di rating*, cit., p. 151 ss. On this point, see also Principe (2014) "Presentazione," in Principe (ed.) *Le Agenzie di Rating* (Milano), p. VII ff., where the author explains the need to clarify the different issues relate to the rating assessments.
45. Regarding the CRAs, the European regulatory framework is actually composed by the Regulation (EU) no. 1060/2009, modified by the Regulation(EU) no. 513/2011 and the Regulation (EU) no. 462/2013, in which the oversight functions are attributed to ESMA. Instead, the Dodd-Frank Act (2010) regulates the credit rating agencies in the American markets, with a specific focus on their conflict of interests and their civil liability; see Acharya et al. (2010) *Regulating Wall Street: The Dodd-Frank Act and the New Architecture of Global Finance* Vol. 608 (Hoboken, NJ); Alcubilla and Ruiz Del Pozo (2012) *Credit Rating Agencies on the Watch List: Analysis of European Regulation* (Oxford).
46. See Coval, Jurek, and Stafford (2009) "Economic Catastrophe Bonds", *The American Economic Review*, Vol. 99, Issue 3, p. 628 ff., where the authors confirm some of the ideas shown in Coval, Jurek, and Stafford (2009) "The Economics of Structured Finance", *The Journal of Economic Perspectives*, Vol. 23, Issue 1, p. 3 ff.; and Masera (2012) "CRAs: Problems and Perspectives," *Analisi giuridica dell'economia*, no. 2, p. 425 ff.

See also Rablen (2013) "Divergence in Credit Ratings", *Finance Research Letters*, Vol.10, Issue 1, p. 12 ff., where the author examines a model in which a CRA operates in both the market for structured products and for corporate debt, and shares a common reputation across the two markets.
47. See Bolton, Freixas, and Shapiro (2012) "The Credit Ratings Game", *Journal of Finance*, Vol. 67, Issue 1, p. 85 ff.
48. See Malmendier and Shanthikumar (2007) "Are Small Investors Naïve About Incentives?", *Working Paper of the Department of Economics Harvard Business School*, UC Berkeley Harvard University, where the authors analyze how investors account for such distortions. Using the NYSE Trades and Quotations database, they found that small traders follow recommendations literally, and then present evidence on the returns of these strategies and discuss possible explanations for the differences in trading response, including informational costs and investor naiveté.
49. According to art. 8d, in case of use of multiple credit rating agencies, the issuer or a related third party shall consider appointing at least one credit rating agency with no more than 10 per cent of the total market share.

6 Non-Standard Operations in the Shadow Banking System

1. See Tucker (2012) *Shadow Banking – Thoughts for a Possible Policy Agenda*, cit., p. 6 ff., where it is highlighted that "these markets are vital to efficient capital markets."

2. See Beber and Pagano (2011) "Short-Selling Bans Around the World: Evidence from the 2007–09 Crisis," *Journal of Finance*, where the authors describe the choice of the global regulators on banning short-selling practices after the beginning of the crisis. This choice causes important variations on short-selling activities that generate various effects on liquidity, price discovery, and stock prices.
3. See Tucker (2012) *Shadow Banking – Thoughts for a Possible Policy Agenda*, cit., p. 6.
4. See ISLA, *Global Master Securities Lending Agreement*, January 2010, p. 3.
5. See Zhang (2014) "Collateral Risk, Repo Rollover and Shadow Banking," *LSE Working Paper*, SSRN no. 2496915 for a dynamic model of the shadow banking system that intermediates funds through the interbank repo market to understand its failing mechanism during the recent financial crisis.
6. See ISLA, *Global Master Securities Lending Agreement*, cit., p. 9.
7. See ISLA, *Global Master Securities Lending Agreement*, cit., p. 10. On this point see also Ali, Ramsay, and Saunders (2013) "The Legal Structure and Regulation of Securities Lending," *CIFR Paper*, no. 022/2014, where the authors outline the regulation of securities lending and short selling, including restrictions on short selling and the applicable disclosure requirements.
8. See ISLA, *Global Master Securities Lending Agreement*, cit., pp. 19–23.
9. See ICMA-SIFMA, *Global Master Repurchase Agreement*, April 2011, art. 1. See also Sakurai and Uchida (2013) "Rehypothecation Dilemma: Impact of Collateral Rehypothecation on Derivative Prices Under Bilateral Counterparty Credit Risk," cit., for a specific analysis of the practice where a derivatives dealer reuses collateral posted from its end user in over-the-counter (OTC) derivatives markets. Moreover, the authors highlight that, although rehypothecation benefits the end user through cost reduction of derivative trades, it also creates additional counterparty credit risk (since the end user may not receive the collateral back when the dealer suddenly defaults).
10. See Tucker (2012) *Shadow Banking – Thoughts for a Possible Policy Agenda*, cit., p. 6.
11. See Tucker (2012) *Shadow Banking – Thoughts for a Possible Policy Agenda*, cit., p. 6, where it is added that "it is also worth mentioning that some asset managers have no – zero – appetite to hold the underlying paper outright in the event of their counterparty defaulting, either because the assets are not covered by investment mandates or the fund managers do not know how to manage them, etc."
12. See Tobias and Shin (2009) *Money, Liquidity and Monetary Policy*, Federal Reserve Bank of New York Staff Report no. 360, *passim*.
13. See Tobias and Shin (2009) *Money, Liquidity and Monetary Policy*, cit., Abstract.
14. See Tobias and Shin (2009) *Money, Liquidity and Monetary Policy*, cit., Abstract.
15. See Constâncio (2012) *Introductory Remarks to the ECB Workshop – Repo Market and Securities Lending: Towards an EU Database*, available at http://www.ecb.europa.eu.
16. See Claessens, Pozsar, Ratnovski, and Singh (2012) *Shadow Banking: Economics and Policy*, cit., p. 22.
17. See Tucker (2012) *Shadow Banking – Thoughts for a Possible Policy Agenda*, cit., p. 7, who recalls the "Securities Lending and Repo: Market Overview and

Financial Stability Issues," and interim report of an FSB group chaired by David Rule.
18. See Claessens, Pozsar, Ratnovski, and Singh (2012) *Shadow Banking: Economics and Policy*, cit., p. 30, where it is summarized the empirical evidence that "around 50 to 70 percent of repo operations in the United States are cleared using TPR, with recent volumes approaching $1.8 trillion, much below the $3 trillion in 2008. The TPR market is a major source of wholesale funding for banks and dealer banks. The U.S. market is serviced by two clearing banks, Bank of New York and JP Morgan, both determined to be systemic by the Financial Stability Board. Pledged collateral is held with custodians and cannot be repledged. The TPR arrangement has several advantages: outsourcing collateral management to the TPR clearer, saving back- office costs for counterparties, and creating economies of scale, as securities are simply moved from one account to another within the clearer's books. It also allows market participants to exchange collateral baskets, outsource risk management (haircut calculation, margin calls, and substitution), pricing, and other ancillary tasks."
19. See Tucker (2012) *Shadow Banking – Thoughts for a Possible Policy Agenda*, cit., p. 7, because he said that "this has become apparent from the Lehman and MFG bankruptcies."
20. It is also closely connected with what was called the process-velocity (in circulating the money), but it is different from the above because—in the banking sector—intermediaries internalize this process (in their assets), while—in the financial market—agents achieve multilateral trade that amplify the possibility of capital circulation.
21. This follows the capacity to which correlates the judgment, clearly formulated in the economic analysis, that the offering must "refer to the enhancement of the credit quality of debt issued by the intermediary"; see Pozsar, Adrian, Ashcraft, and Boesky (2012) *Federal Reserve Bank of New York Staff Reports – Shadow Banking*, cit., p. 5.
22. See Draghi (1997) "Commento sub art. 46 d. lgs. 415 del 1996," in Capriglione (ed.), *La disciplina degli intermediari e dei mercati finanziari*, (Padova), p. 385.
23. We refer to the kind of problems raised by the *Statement by European Commissioner for Competition Margrethe Vestager on Tax State Aid Investigations*, cit, November 6, 2014.
24. On this point see Greene and Broomfield (2013) "Promoting Risk Mitigation, Not Migration: A Comparative Analysis of Shadow Banking Reforms by the FSB, USA and EU," *Capital Markets Law Journal*, vol. 8, no. 1, cit., where the authors outline the need for an increased analysis (by the supervisors) of shadow banking activities (instead of entity-based strategies, imposing bank-like regulation). This suggest the possibilities of a more effective identification of the risks' sources, a better uniformity in cross-border application of reforms, and to a greater flexibility in addressing financial innovation.
25. See Gorton and Souleles (2005) "Special Purpose Vehicles and Securitization," cit., where it is highlighted that SPVs become sustainable in a repeated SPV game, because credit institutions can implicitly commit to subsidize or bail out their SPVs when these would not honor their debts.

26. See Coval, Jurek, and Stafford (2008) "The Economics of Structured Finance," *Harvard Business School Finance Working Paper*, no. 09-060, cit., where the authors highlight the essence of structured finance activities in the pooling of economic assets (such as loans, bonds, mortgages) and succeeding issuance of a tranche (prioritized capital structure of claims), against these collateral pools. The authors show, as a result of this prioritization scheme, that many of the tranches are far safer than the average asset in the underlying pool.
27. See Lener and Lucantoni (2012) "Regole di condotta nella negoziazione degli strumenti finanziari complessi: disclosure in merito agli elementi strutturali o sterilizzazione, sul piano funzionale, del rischio come elemento tipologico e/o normativo?" *Banca borsa e titoli di credito*, I, p. 369 ff.
28. See Tucci (2014) "'Interest Rate Swaps': 'causa tipica' e 'causa concreta'," *Banca borsa e titoli di credito*, II, p. 291 ff.; Tucci (2013) "La negoziazione degli strumenti finanziari derivati e il problema della causa del contratto," *Banca borsa e titoli di credito*, I, p. 68 ff.,on the problem of the Italian concept of contractual "causa," useful also for the analysis of the "consideration" of any swap.
29. See Tian (2011) *Shadow Banking System, Derivatives and Liquidity Risks*, MFA Annual Meeting, Chicago.
30. See Sakurai and Uchida (2013) "Rehypothecation Dilemma: Impact of Collateral Rehypothecation on Derivative Prices Under Bilateral Counterparty Credit Risk," cit., where there is an evaluation of the benefits and risks of derivatives rehypothecation. In particular, the proposal of a "derivative pricing framework" takes into account the presence of a bilateral counterparty credit risk, which should determine the rehypothecable collateral's amount.
31. It is useful to recall the analysis of Sakurai and Uchida (2013) "Rehypothecation Dilemma: Impact of Collateral Rehypothecation on Derivative Prices Under Bilateral Counterparty Credit Risk," cit., where there is a model in which two different types of collateral are used: the time delay of collateral posting and the rating-dependent collateral agreement.
32. See Classens, Pozsar, Ratnovsky, and Singh (2012) *Shadow Banking: Economic and Policies*, cit., Appendix 1. Over-the-Counter Derivatives: Central Counterparties and Under-Collateralization, p. 29
33. See ISDA (2014) *ISDA Comments – EU proposal on Structural Reform of the EU Banking Sector*, July, 2, p. 5.
34. See Piga (2001) "Do Governments use Financial Derivatives Appropriately? Evidence from Sovereign Borrowers in Developed Economies," *International Finance*, p. 189 ff.
35. It goes without saying that Greece is an example, and the Grexit is a risk that the shadow banking system cannot avoid; see Athanassiou (2009) "Withdrawal and Expulsion from the EU and EMU. Some Reflections," *Legal Working Paper* of ECB, p. 39 ff.
36. For an analysis of the *European building blocks*, see Capriglione (2013) *L'Unione bancaria europea*, cit., p. 111.
37. See Lemma and Haider (2012) "The Difficult Journey Towards European Political Union: Germany's Strategic Role," *Law and Economics Yearly Review*, p. 390 ff., where there is the following conclusion: "it remains the necessity to define new development trends of the European legal system, which

currently is hard-fought between the methodological austerity of the public finance (as framed since the Maastricht Treaty) and the auspices of a subsidiarian solidarism of the EU institutions (promoted by Mediterranean instances)."
38. See Montedoro (2010) *Mercato e potere amministrativo*, (Napoli) p. 13 and p. 319.
39. Illustrative in this regard is the diversity of ways in which the financial crisis occurred in Europe. Indeed, with regard to the effects of the latter on sovereign debt, it can be observed that Italy has failed to safeguard its balance by resorting to taxation, whereas other countries have instead requested external aid to its economy (i.e. Greece, Ireland, Portugal, etc.). It should be noted that—even if it had been adopted in the EMU (from the beginning), an austerity and rigor-oriented policy—there would not have been a solution to the structural problems, from ancient times, levied on Mediterranean countries (i.e. high levels of public debt, low tax revenue). This is because it seems plausible to assume that the cause of one of the above imbalances lies in the shortness of the time period granted to the states taking part to the monetary union to implement the economic and legal convergences necessary for a full harmonization. This suggests a possible justification for use, by the member states, of the negotiation of derivative financial instruments able to reschedule the financial burden (i.e. able to restructure public debt both in its interest and principal amount).
40. As a result, there is the need to verify the validity of the safeguards put in place by the system, to face the dangers of certain over-the-counter operations, privately concluded between states and international investment banks. The regulation must avoid these operations being resolved to the detriment of the states. In the absence of such safeguards, in fact, the Treasury would be exposed to the risks that—as we have seen—characterize the system under observation.
41. See Awrey (2010) *The Dynamics of OTC Derivatives Regulation: Bridging the Public-Private Divide*, in *Oxford Legal Studies Research Paper*, where the author "explores both the private and social costs and benefits of OTC derivatives and the respective strengths and weaknesses of public and private systems of ordering in pursuit of the optimal mode of regulating OTC derivatives markets."
42. See Capriglione, *Intermediari finanziari investitori mercati*, cit., p. 214–215.
43. In this context there is also the trouble of the unsolicited ratings on sovereign debts, see Troisi (2013) *Le Agenzie di Rating*, cit., p. 87 ff.
44. It is not rare to see this question in the press, see Frisone (2013) "Le insidie degli swap plain vanilla," *Il Sole 24 Ore*, November 9, p. 24.
45. See Anelli (1998) "La responsabilità risarcitoria delle banche per illeciti commessi nell'erogazione del credito," cit., p. 137 ff.
46. See, on this point, Lemma (2006) "L'applicazione del Fair Value alle banche: problematiche giuridiche e soluzioni," *Banca borsa e titoli di credito*, I, p. 723 ff.
47. See Monti (2000) *Manuale di finanza per l'impresa*, (Torino) p. 146, for an analysis of the forms in which these operations were designed in the beginning of the third millennium.

48. For a pre-crisis perspective, see Barth, Caprio, and Levine (2001) "The Regulation and Supervision of Banks Around the World: A New Database," *World Bank Policy Research Working Paper*, no. 2588.
49. See art. 449 "Exposure to Securitisation Positions" of the Regulation (EU) no. 575/2013.

7 Shadow Banking Risks and Key Vulnerabilities

1. Consequently, the creation of value made by the shadow processes is the profit of the arrangers, which is vulnerable by the effects of specific risks.
2. See *Report of the High-Level Group on Financial Supervision in the EU*, chaired by Jacques De La Rosiere, Brussels, February 25, 2009, p. 11.
    See also Troeger (2014) "How Special Are They? – Targeting Systemic Risk by Regulating Shadow Banking," where the author shows that the general policy goals of prudential regulation should legitimize an intervention in the shadow banking sector. In particular, the author concludes that a stricter regulatory treatment of financial innovation shall reduce the circumventions of existing rules and standards.
3. See Langfield and Pagano (2015) "Bank Bias in Europe: Effects on Systemic Risk and Growth," *ECB Working Paper*, no. 1797, where the authors argue that the phenomena arise owing to an amplification mechanism, by which banks overextend and misallocate credit when asset prices rise, and ration it when they drop, and then conclude by discussing policy solutions to Europe's "bank bias," which include reducing regulatory favouritism towards banks, while simultaneously supporting the development of securities markets; see also Espinosa, Vega, and Russell (2015) "Interconnectedness, Systemic Crises and Recessions," *IMF WorkingPaper*, no. 15/46 where the authors attempt to capture and integrate four widely held views about financial crises.
4. It must be considered that the regulation of this phenomenon identifies a significant moment in the process of evolution of the international regulatory framework, that—since the 1980s—the Basel Committee on Banking Supervision—acting as a forum for regular cooperation on banking supervisory matters—has brought deep transformations to the regulation of capital intermediation.
5. See Capriglione (1994) *L'ordinamento finanziario verso la neutralità* (Padova), p. VII.
6. This regulatory process has arrived to the "International Regulatory Framework for Banks," prepared by the Basel Committee on Banking Supervision (in June 2011), which introduces a comprehensive set of reform measures to strengthen the regulation, supervision, and risk management of the banking sector.
7. See Tyson and Shabani (2013) "Sizing the European Shadow Banking System: A New Methodology," cit., where there is the total amount of £ 900 billion.
8. See Knight (1921) *Risk, Uncertainty, and Profit*(Cambridge), where it is highlighted that "the difficulties...arisen from a confusion of ideas which goes deep down into the foundations of our thinking."
9. See Acharya, Schnabl, and Suarez (2012) "Securitization Without Risk Transfer," *CEPR Discussion Paper*, no. DP8769, where the authors analyze

asset-backed commercial paper conduits, which experienced a shadow-banking "run" and played a central role in the early phase of the financial crisis of 2007–2009.
10. This is a dynamic shown by the recent crisis, see Anand Sinha (2013) *Regulation of Shadow Banking – Issues and Challenge*, cit., where it is stated that "these effects were powerfully revealed during the global financial crisis in the form of dislocation of asset-backed commercial paper (ABCP) markets, the failure of an originate-to-distribute model employing structured investment vehicles (SIVs) and conduits, 'runs' on MMFs and a sudden reappraisal of the terms on which securities lending and repos were conducted."
11. See Moreira and Savov (2014) "The Macroeconomics of Shadow Banking," *NBER Working Paper*, no. w20335, for an interesting macroeconomic model that centers on liquidity transformation in the financial sector. It shall be useful to take into account the idea that intermediaries can maximize liquidity creation by issuing securities that are money-like in normal times but become illiquid in a crash when collateral is scarce. Even if these authors call this process shadow banking, it can be considered that this is only a kind of shadow banking operations.
12. See O'Doherty, Savin, and Tiwari (2011) "Modeling the Cross Section of Stock Returns: A Model Pooling Approach," *Journal of Financial and Quantitative Analysis (JFQA)*, where the authors find out how the benefits to model pooling are most pronounced during periods of such economic distress and recognize it as a valuable tool for asset allocation strategies.
13. See Briand, Nielsen, and Stefek (2009) "Portfolio of Risk Premia: A New Approach to Diversification," *MSCI Barra Research Paper*, no. 2009-01, where the authors confirm the benefits of diversification (in terms of less volatility) with a simple asset allocation case study, by comparing a 60/40 equity/fixed income allocation with an equal weighted allocation across eleven style and strategy risk premia.
14. See Anand Sinha (2013) *Regulation of Shadow Banking – Issues and Challenge*, cit.; see also the position of European Commission which said that "the 2008 crisis was global and financial services were at its heart, revealing inadequacies including regulatory gaps, ineffective supervision, opaque markets and overly-complex products. The response has been international and coordinated through the G20 and the Financial Stability Board (FSB)"; see European Commission, *Green Paper on Shadow Banking*, cit., p. 2.
15. See FSB, *Financial Reforms – Update on Progress to G20 Finance Ministers and Central Bank Governors*, April 4, 2014, p. 1.
16. See European Commission, *Communication from the Commission to the Council and the European Parliament, Shadow Banking – Addressing New Sources of Risk in the Financial Sector*, cit., pp. 3–4.
17. This is a reality analyzed by the European Commission that shows the negative externalities of a system qualified both by a cross-jurisdictional reach and an inherent mobility of securities and fund markets, see European Commission, *Green Paper on Shadow Banking – Frequently Asked Questions*, cit.

See also Maimeri (2011) "Criteri di proporzionalità ed efficacia dei modelli di risk management," *Diritto della banca e del mercato finanziario*, f. 2, pt. 1, p. 241 ff., for an analysis of the Italian legislation on this topic.

18. See Adrian, Ashcraft, and Cetorelli (2013) "Shadow Bank Monitoring," *FRB of New York Staff Report*, no. 638, where the authors, by describing the characteristics of the shadow banking system and its interconnectedness with regulated financial institutions, call for a massive action of monitoring risks taking into account the recent efforts by the FSB.
19. See Campbel and Kracaw (1980) "Information Production, Market Signalling, and the Theory of Financial Intermediation," *The Journal of Finance*, Vol. 35, Issue 4, p. 863 ff.
20. It is, therefore, necessary to take regulatoryaction (of "hard law") that, avoiding outright bans, addresses the issues of market-based finance, according to the criteria of equity and efficiency.
21. See, on this topic, Antonucci (2009) "Regole di condotta e conflitti di interesse," *Banca borsa e titoli di credito*, 2009, f. 1, p. 9 ff.
22. See Luttrell, Rosenblum, and Thies (2012) "Understanding the Risks Inherent in Shadow Banking: A Primer and Practical Lessons Learned," *Staff Papers Federal Reserve Bank of Dallas*, no. 18.
23. See Bloise, Reichlin, and Tirelli (2013) "Fragility of Competitive Equilibrium with Risk of Default," *Review of Economic Dynamics*, p. 271 ff.
24. See Jobst (2010) "The Credit Crisis and Operational Risk—Implications for Practitioners and Regulators," *Journal of Operational Risk*, Vol. 5, Issue 2, where the author highlights the increased importance of operational risk underneath greater systemic risk concerns. In fact, from the author's point of view, the fallout from the financial crisis has demonstrated that many sources of systemic risk were triggered by vulnerabilities in operational risk management—which has not kept pace with financial innovation—and that there has been an excessive focus of regulation on prudential requirements, with a lack of identification of substantial operational risk in market-based liquidity transformation.
25. On the fees used in the practices, see Sciarrone Alibrandi (2011) "Le clausole di remunerazione degli affidamenti," *Analisi Giuridica dell'Economia*, f. 1, p. 169 ff.
26. One must take into account also the "traditional banking backstops," which, in the last century, was used to manage the banking risks; see Capriglione (2012) "Commento sub art. 5 d. lgs. 385 del 1993," cit., where the analysis of the supervisory goals allows to understand that only certain technical solution can be applied to the entities involved in "Shadow credit intermediation process." See also Luttrell, Rosenblum, and Thies (2012) "Understanding the Risks Inherent in Shadow Banking: A Primer and Practical Lessons Learned," cit., who highlight that "Any Financial institution that takes on maturity transformation faces the possibility of a running large numbers of depositors simultaneously demanding their funds, typically at a time of panic. Bank runs were commonplace in the nineteenth and early twentieth centuries, with thousands of banks failing in the 1920s and early 1930s alone."

See also Acharya, Schnabl, and Suarez (2012) "Securitization Without Risk Transfer," cit., where the analysis of asset-backed commercial paper conduits shows the experienced shadow-banking runs, and their central role in the early phase of the financial crisis of 2007 to 2009. This analysis shows that regulatory arbitrage was one of the main motive behind setting up

conduits. An this, because—according to these Authors—the guarantees were structured so as to reduce regulatory capital requirements.
27. See EC Commission, *Green Paper on Shadow Banking – Frequently asked questions*, cit.
28. We shall consider that the risks of an SPV are transmitted according to different paths, other than those abovementioned. We refer in particular to the contracts entered into with third parties in order to obtain liquidity guarantees or credit insurance. Moreover, at the economic level, it was alleged that "credit enhancement can come in the form of overcollateralization or external enhancement," hence the ability of the vehicle to manage its default risk (holding assets of higher value to that of the financial instruments issued) or to transfer it to the banks (which have granted credit lines) or insurance companies (which have subscribed a policy); see Luttrell, Rosenblum, and Thies (2012) "Understanding the Risks Inherent in Shadow Banking: A Primer and Practical Lessons Learned," cit., p. 25,
29. See Luttrell, Rosenblum, and Thies (2012) "Understanding the Risks Inherent in Shadow Banking: A Primer and Practical Lessons Learned," cit., p. 25
30. Penalizing, in the measurement of risk governance, it is the lack of data and news that can indicate, in detail, the actual reliability of the entities and their activities (hardly ponderable also in relation to credit, market and operational risks). In addition, there is the opacity of the evaluations of the risk of liquidity (in the short-term) and structural imbalances (in the long-term), to which is associated the lack of disclosure requirements on the financial leverage (including off balance sheet exposures). Consequently, it appears to be difficult to assess the overall risk exposure of the governance of the "shadow banking entities," whereas the absence of regulatory safeguards (acting as backstop) and patrimonial requirements adversely affects the ability to predict their systemic growth.
31. It is helpful to consider the "Recognition traditional securitisation of significant risk transfer" (according to art. 243, par. 4, and 244, par. 4 CRR), and to the use additional own funds requirements for securitisations of revolving exposures (art 256, par. 7 CRR), and to the calculation of risk-weighted exposure amounts under the IRB approach (or an Internal Assessment Approach, art. 259, par. 1 lett. c CRR) or with the *"look-through methodology"* (art. 259, par. 1, and par. 3 CRR).
32. We agree with Luttrell, Rosenblum, and Thies (2012) "Understanding the Risks Inherent in Shadow Banking: A Primer and Practical Lessons Learned," cit., p. 37 where it is highlighted that "to identify the major players in today's interwoven global financial-intermediation landscape, the FSB created a list of G-SIBs in 2011 and began in November to update it annually".
33. See Luttrell, Rosenblum, and Thies (2012) "Understanding the Risks Inherent in Shadow Banking: A Primer and Practical Lessons Learned," cit., pp. 38–39.
34. See Benigno and Romei (2014) "Debt Deleveraging and The Exchange Rate," *Journal of International Economics*, 93, p. 1 ff.
35. See Rule (2012), "Collateral management in central bank policy operations," in *Bank of England, Centre for Central Banking Studies Handbook*, no. 31 about the criteria used for collateralized lending of central banks, with regard to ECB and FED.

36. See Adrian and Liang (2014) "Monetary Policy, Financial Conditions, and Financial Stability," cit., for an analysis of monetary policy transmission channels and financial frictions that give rise to this trade-off between financial conditions and financial stability.
37. See Goodhart (2005) "Financial Regulation, Credit Risk and Financial Stability," *National Institute Economic Review*, Vol. 192, Issue 1, p. 118 ff.
38. See Adrian and Shin (2009) *The Shadow Banking System: Implications for Financial Regulation*, no. 382, Staff Report, Federal Reserve Bank of New York.
39. See Luttrell, Rosenblum, and Thies (2012) "Understanding the Risks Inherent in Shadow Banking: A Primer and Practical Lessons Learned," cit., p. 23.
40. See Lemma (2013) "Informazione finanziaria e tutela dei risparmiatori," in Capriglione (ed.), *I contratti del risparmiatore* (Milano), p. 259 ff.
41. See European Commission, *Green Paper on Shadow Banking – Frequently Asked Questions*, cit., where it is clarified that "shadow banking operations can be used to avoid regulation or supervision applied to regular banks by breaking the traditional credit intermediation process in legally independent structures dealing with each other. This 'regulatory fragmentation' creates the risk of a regulatory 'race to the bottom' for the financial system as a whole, as banks and other financial intermediaries try to mimic shadow banking entities or push certain operations into entities outside the scope of their consolidation. For example, operations circumventing capital and accounting rules and transferring risks outside the scope of banking supervision played an important role in the build-up to the 2007/2008 crisis."
42. This measure does not propose to deal with lending to households for the purposes of house purchases. Eligible lending to the non-financial private sector in the context of this measure thus excludes loans to households for the purpose of house purchases; see Decision of the European Central Bank of 29 July 2014 on measures relating to targeted longer-term refinancing operations (ECB/2014/34), para 2.
43. See BIS, *Central Bank Collateral Frameworks and Practices*. A report by a study group established by the Markets Committee (study group chaired by Guy Debelle, Assistant Governor of the Reserve Bank of Australia), March 2013, p. 5.
44. See BIS, *Central Bank Collateral Frameworks and Practices*, cit., p. 5.
45. See BIS, *Central Bank Collateral Frameworks and Practices*, cit., pp. 20–32, where it is highlighted that "as the ability to obtain market funding wanes or even vanishes, counterparties have a greater need to access central bank liquidity. In this environment, the opportunity cost of bringing less liquid assets to the central bank may fall, given that these assets are no longer as valuable in the market," whereas "in less volatile times, however, it seems that wider eligibility does not necessarily always attract less liquid collateral." Moreover, consider that, "for example, in Switzerland, more than 99% of all repo transactions in the prevailing interbank repo market are covered by SNB-eligible collateral."
46. See Bank of England, *Liquidity Insurance at the Bank of England: Developments in the Sterling Monetary Framework*, October 2013, p. 1, 4th edition.
47. See Benigno and Paciello (2014) "Monetary Policy, Doubts and Asset Prices, *Journal of Monetary Economics,* 64, p. 85 ff.

8 The Shadow Banking System and the Need for Supervision

1. Moreover, in Europe, the institutional model went beyond the public nature of stock exchanges and the pervasive presence of supervisors in the business of banking. Indeed, already Directives nos. 93/6/EEC and 93/22/EEC had clearly indicated the independent purposeful and propulsive activity of brokers and managers; see Capriglione (1997) "Presentazione," in Capriglione (ed.) *La disciplina degli intermediari e dei mercati finanziari*, (Padova) p. XIV.
2. See Draghi (1997) "Commento sub art. 46 d. lgs. 415 del 1996," in Capriglione (ed.) *La disciplina degli intermediari e dei mercati finanziari*, (Padova) p. 384 ff., where the author shows the guidelines of the reform process aimed by directive 93/22/EC and the new roles of the financial markets as private enterprises able to self-regulate their activities, within a competitive system.

    See also Draghi (2009) "Guido Carli innovatore," Speech at Accademia Nazionaledei Lincei, 16 gennaio 2009; Draghi (2010) "Modernisation of the Global Financial Architecture: Global Financial Stability," Remarks of the Chairman of the Financial Stability Board to the Committee on Economic and Monetary Affairs European Parliament, March 17, 2010, where the above ideas are referred to the recent financial crisis.
3. See Valentino (1996) "Decreto Eurosim: focus sul big bang del mercato finanziario italiano," *Le Società*, n. 9, p. 1006 ff.
4. See Draghi (1997) "Commento sub art. 46 d. lgs. 415 del 1996," cit., p. 389, where the author emphasizes the risk that the regulatory strategy was aimed to promote the "old regime" than new market as enterprises.
5. See Sepe (2000) *Il risparmio gestito*, (Bari) p. 31 ff.
6. See Posner (1993) "The New Institutional Economics Meets Law and Economics," *Journal of Institutional and Theoretical Economics*, Vol. 149, Issue 1, p. 73 ff.
7. In that way, a technique for cross-border operation was found, and it required the adaptation of markets to competitive criteria (due to the possibility for operators to transfer trade in the financial markets more convenient in terms of operational possibilities, quality and price); seePadoaSchioppa (1988) "Verso un ordinamento bancario europeo," *Bollettino economico della Banca d'Italia*, no. 10. p. 57 ff., where the author links the need for a cross-border opening to the contents of the Single European Act (SEA). See also Costi (1991) "La seconda direttiva di coordinamento: i principi," in Cesarini and Scotti Camizzi (eds) *Le direttive della CEE in material bancaria* (Milano) p. 67 ff.; Capriglione and Sepe (1997) "Banche estere operanti in Italia," *Enciclopedia giuridica Treccani* (Roma), IV, p. 12 ff.
8. See Draghi (1997) "Commento sub art. 46 d. lgs. 415 del 1996," cit., p. 388.
9. See Enriques (2005) "Conflicts of Interest in Investment Services: The Price and Uncertain Impact of MiFID's Regulatory Framework," *University of Bologna and ECGI Working Paper*, available at SSRN no. 782828, where the author predicted "that, as a consequence of the home country control principle (investments firms and banks will have to comply with their own home country rules wherever they conduct their business within the EU),

the MiFID will lead to *de facto* complete harmonisation of the rules on conflicts of interest: Member States will refrain from enacting further rules on conflicts of interest so as to avoid to put domestic firms at a competitive disadvantage vis-à-vis EC firms."

10. See Recital no. 61, Directive no. 2014/65/CE; see also Kirilenko, Kyle, Samadi, and Tuzun (2014) "The Flash Crash: The Impact of High Frequency Trading on an Electronic Market," *SSRN* Paper, no. 1686004, where the authors try to understand if, in a particular case, HFTs cause certain crash events or they can exacerbate the price movement by absorbing immediacy ahead of others.
11. See Gorton (2009) "Slapped in the Face by the Invisible Hand: Banking and the Panic of 2007," *SSRN Working Paper*, no. 1401882, p. 1, where the author admits that "many private decisions were made, over a long time, which created the shadow banking system."
12. See Cioffi (2010) "Persistence and Perversity: The Global Financial Crisis, the Failure of Reform, and the Legitimacy Crisis of Finance Capitalism," *Western Political Science Association 2010 Annual Meeting Paper*, for a wider analysis of this topic from a political science perspective.
13. See FSB (2014) *FSB Reports to the G20 on Progress in Reforming Resolution Regimes and Resolution Planning*, November 12, 2014, and its *2014 Global Shadow Banking Monitoring Report*, cit, p. 35, where it is announced that the FSB, through its workstream, will try to identify the relevant authorities with oversight of shadow entities, and then revision of availability of policy tools to address the identified risks.
14. See Hobsbawm (1995) *Age of Extremes. The Short Twentieth Century*, cit.; Capriglione (2010) "Un secolo di regolazione," in Capriglione (ed.) *L'ordinamento finanziario italiano*, cit., p. 53 ss.
15. See Demirgüç, Kunt, and Huizinga (1999) "Market Discipline and Financial Safety Net Design," *World Bank Policy Research Working Paper*, no. 2183, where the authors outline that "the safety net that policymakers design must provide the right mix of market and regulatory discipline—enough to protect depositors without unduly undermining market discipline on banks."
16. For the juridical fundaments of such a decision see Oppo (1990) "Commento sub art. 41 Cost," Capriglione and Mezzacapo (eds) *Codice commentato della banca*, (Milano) I, p. 3 ff.; Merusi (1980) "Commento sub art. 47 Cost.," Branca (ed.), *Commentario alla Costituzione* (Bologna-Roma), III.
17. See Posner (2007) *Economic Analysis of Law*, (New York) p. 419 ff. and p. 465 ff., for a closer look at the market for corporate securities, in IRS double dimension: risk and expected return.
18. See Vento (2011) "Shadow Banking and Systemic Risk: In search for Regulatory Solutions," paper presented at European Association of University Teachers of Banking and Finance Annual Seminar, Valencia, Spain.
19. To the hypothesized renounce of the authority to impose a constraint-system (on alternative markets) is added, therefore, also the realization of the limited extension (and strength) of the national systems (confined within the territorial boundaries of the States). This reflects the ideas of Montedoro and Supino (2014) "Il difficile dialogo fra economia

e diritto in una prospettiva istituzionalistica," *Rivista Trimestrale di Diritto dell'Economia*, p. 135 ff.
20. See Lemma (2013) *Etica e professionalità bancaria*, cit., p. 134 and, in line with the foregoing, see Solomon (2000) "Historicism, Communitarianism and commerce: An Aristotelean Approach to Business Ethics," *Contemporary Economic Ethics and Business Ethics* (Berlin) p. 119; Black (2001) "The Legal and Institutional Preconditions for Strong Securities Markets," *UCLA Law Review*, p. 781 ff.; Koslowski (2001) *Principles of Ethical Economy* (Dordrecht-Boston-London), p. 13; Latouche (2003) *Giustizia senza limiti. La sfida dell'etica in un'economia mondializzata* (Torino) p. 254 ss.; Stirner (2006) *The Ego and its Own* (Cambridge) p. 141 ff.; Wilmarth (2009) "The Dark Side of Universal Banking: Financial Conglomerates and the Origins of the Subprime Financial Crisis," *Connecticut Law Review*, vol. 41, no. 4.
21. See Morera and Rangone (2013) "Sistema regolatorio e crisi economica," *Analisi Giuridica dell'Economia*, fasc. Vol. 2, p. 383 ff., for a focus on the *Italian* regulation and crisis.
22. See Renne (2014) "Options Embedded in ECB Targeted Refinancing Operations," *Banque de France Working Paper*, no. 518, for a first analysis of the implementation of new refinancing operations aimed at supporting bank lending to the non-financial private sector (announced in June 2014). It is important to recall this paper because it focuses on the options embedded in these targeted longer-term refinancing operations. In particular, it shows how quantitative results point to substantial gains—for participating banks—attached to the satisfaction of lending conditions defined by the scheme.
23. See Mésonnier and Monks (2014) "Did the EBA Capital Exercise Cause a Credit Crunch in the Euro Area?," *Banque de France Working Paper*, no. 491.
24. And I shall highlight how this is helping in the introduction of the European System of Financial Supervision (so called ESFS), as well as the new supervisory policies (which ended, within the euro zone, into the constitution of the European Banking Union, so called EBU); see Atik (2014) "EU Implementation of Basel III in the Shadow of Euro Crisis," *Review of Banking and Financial Law*, vol. 38, p. 287, where the author—after having the possibility of selective adoption of new prudential rules—discusses the future of Basel III in the EU as the euro zone approaches a banking union.
25. See the case of *Bank of Credit and Commerce International* and the recent events of banking default; See Stiglitz (2010) *Freefall*, New York, p. 238 ff. and the "Plan de soutien aux banques," in *Intervention de M. le Président de la République*; as well as the "Special Resolution Regime" of the British "Banking Act," of February 12, 2009.

See also *Financial Stability and Depositor Protection: Strengthening the Framework*, edited by BoE, HMT and FSA, Norwich, 2008.
26. See Enriques and Zetzsche (2014) "Quack Corporate Governance, Round III? Bank Board Regulation Under the New European Capital Requirement Directive," *Oxford Legal Studies Research Paper*, no. 67/2014, where the Authors argue that European policymakers and supervisors should avoid using a heavy hand, respectively when issuing rules implementing CRD IV provisions on bank boards and when enforcing them.

27. See Masera (2014) "CRR/CRD IV: The Trees and the Forest," *SSRN Working Paper*, no. 2418215, where the author reviews the new CRR/CRD IV capital regulatory framework and highlights the weaknesses that continue to characterise the new capital regulatory framework.
28. See Turner (2012) "Shadow Banking and Financial Instability," *The Harvard Law School Forum*, April 16, 2012 where it is highlighted that "because of these...features, entirely free banking systems—such as existed in the USA before the creation of the Federal Reserve in 1913—were inherently unstable".
29. In this context, the EC, *Communication on Shadow Banking and Proposal on Money Market Funds*, September 4, 2013 is the basis for a new action plan aimed to restore sustainable health and stability to this sector by addressing the shortcomings and weaknesses highlighted by the crisis. This also explain why the Commission's approach regarding the shadow banking sector consists of delivering transparent and resilient market-based financing, while tackling major financial risks. Despite this, the risk of National will to admit regulatory arbitrages would greatly undermine the future of this proposal.
30. See Capriglione and Troisi (2014) *L'ordinamento finanziario dell'UE dopo la crisi*, cit., p. X.
31. See Ferrarini and Chiarella (2013) "Common Banking Supervision in the Eurozone: Strengths and Weaknesses," *ECGI—Law Working Paper*, no. 223/2013, where the authors highlight that certain the weaknesses of the ESFS could be tempered by an extension of the Banking Union to a sufficient number of non-euro countries under the regime of close cooperation.

    See also Ferran (2014) "European Banking Union and the EU Single Financial Market: More Differentiated Integration, or Disintegration?," *University of Cambridge Faculty of Law Research Paper*, no. 29/2014, for the early indication that some non-euro member states do want to join EBU notwithstanding certain lingering differences between their position and that of euro area member states provide an encouraging sign as to the value of those efforts.
32. See Eurogroup (2014) *Work Programme for the Eurogroup for the Second Half of 2014*, Brussels, June 19, 2014, p. 2, for the policy priorities of this institutions, which shall pay close attention to financial stability in the euro area as well as to the euro area aspects in establishing and operationalizing the Banking Union. It is clear, then, that Eurogroup will follow closely the comprehensive assessment and the stress tests, discuss its implications and certain aspects of the functioning of the SRM and SRF.
33. See EC, *The Juncker Commission: A Strong and Experienced Team Standing for Change*, Brussels, IP/14/984,September 10, 2014.
34. See Rixen (2013) "Why Reregulation after the Crisis is Feeble: Shadow Banking, Offshore Financial Centers and Jurisdictional Competition," cit.
35. See *Opinion of the European Economic and Social Committee Communication from the Commission to the Council and the European Parliament — Shadow Banking — Addressing New Sources of Risk in the Financial Sector*, COM(2013) 614 final, 2014/C 170/09, where it is written that "it notes delay and recommends intensifying and accelerating action to clarify the crucial issue of strengthening oversight of the shadow banking sector, where mention

is simply made of a few issues and the only future measure referred to is the review of the European System of Financial Supervisors (ESFS), to be performed by the Commission in 2013."
36. Cfr. European Commission, *Programme of the Conference "Towards a Better Regulation of the Shadow Banking System,"* April 27, 2012.; see also Vento (2011) "Shadow Banking and Systemic Risk: In search for Regulatory Solutions," cit.
37. See Rixen (2013) "Why Reregulation after the Crisis is Feeble: Shadow Banking, Offshore Financial Centers and Jurisdictional Competition," cit., for a criticism against the lack of regulation and oversight in the shadow banking sector, still too unsupervised after the financial crisis.
38. See Hill (2014) *Capital Markets Union – Finance Serving the Economy*, Brussels, November 6, 2014.
39. See Enria (2009) "The Development of Financial Regulation and Supervision in Europe," in Mayes, Pringle, and Taylor (eds), *Towards a New Framework for Financial Stability* (London), p. 59 ff.
40. See Pellegrini (2012) "Conclusioni," in Pellegrini (ed.), *Elementi di diritto pubblico dell'economia*, (Padova) p. 569.
41. See ESRB, *Annual Report 2013*, p. 8, where there is a subdued economic outlook: challenges facing a banking sector that is still fragile.
42. See Calabresi (1985) *Ideals, Beliefs, Attitudes, and the Law*, (Syracuse), p. 84, where the author cites his (1982) "The New Economic Analysis of Law," *Proceedings of the British Academy*.
43. See Ferran and Alexander (2010) "Can Soft Law Bodies be Effective? Soft Systemic Risk Oversight Bodies and the Special Case of the European Systemic Risk Board," *University of Cambridge Faculty of Law Research Paper*, no. 36/2011, where the authors outline that "strengthening and reinforcing are words that feature prominently in many policy statements relating to these institutional developments but many of these bodies, including the FSB and the ESRB, are designed to operate without legally-binding powers."
44. See Draghi (2013) "Foreword," in ESRB, *Annual Report 2013*, cit., p. 5.
45. See Pellegrini (2012) "L'Architettura di vertice dell'ordinamento finanziario europeo: funzioni e limiti della supervisione," *Rivista Trimestrale di Diritto dell' Economia*, p. 52 ff., where the author clarifies the distinction between macro-prudential supervision, assigned to the ESRB, and a micro-prudential supervision, delegated to the ESFS, a network of national authorities which cooperate with three new European authorities (EBA, ESMA, EIOPA).
46. See Troiano (2010) "L'architettura di vertice dell'ordinamento finanziario europeo," in Pellegrini (ed.), *Elementi di diritto pubblico dell'economia*, cit., p. 542.
47. The duty of sincere cooperation is provided in art. 1, para 4, Regulation (EU) no. 1092 of 2010 (for ESRB); See also art.2, para. 4, Regulation (EU) no. 1093 of 2010 (for EBA); art. 2, para. 4, Regulation (EU) no. 1094 of 2010 (for EIOPA); art. 2, para. 4, Regulation (EU) no. 1095 of 2010 (for ESMA).

For further clarification, see Casolari (2012) "The Principle of Loyal Co-operation: A 'Master Key' for EU External Representation?," in Blockmans and Wessel (eds) *Principles and Practices of EU External Representation*, (The Hague), p. 11 ff.

48. See European Commission, *Green Paper on Shadow Banking*, cit., p. 6; on this point, see also Moreira and Savov (2014) "The Macroeconomics of Shadow Banking," *NBER Working Paper*, no. w20335, for an interesting macroeconomic model, useful to understand the importance of a global perspective in regulating the shadow banking.
49. See Enria, Angelini, Neri, Quagliariello, and Panetta (2010) "Pro-Cyclicality of Capital Regulation: s it a Problem? How to Fix it?", *Bancad'Italia Occasional Paper*, no. 74, where the authors use a macroeconomic euro area model with a bank sector to study the pro-cyclical effect of the capital regulation, and reach the conclusion that "a permanent increase in the capital requirement would have negative consequences on welfare" (p. 38).
50. See European Commission, *Green Paper on Shadow Banking*, cit., p. 6; this approach has been confirmed by FSB (2014), *Global Shadow Banking Monitoring Report*, cit., presenting the data from 25 jurisdictions and the euro area as a whole, covering about 80 per cent of global GDP and 90 per cent of global financial system assets.
51. See Schoenmaker (2012) "Banking Supervision and Resolution: The European Dimension," *Law and Financial Markets Review*, Vol. 6, 2012, p. 52 ff., where the author highlights that "National authorities aim for the least-cost solution for domestic taxpayers. This results in an undersupply of the public good of global financial stability."
52. According to European Commission, *Green Paper on Shadow Banking*, cit., one must take into account that the EU institutions have taken important actions to address shadow banking issues raised by certain securitization structures, with the effect to avoid that the shadow operations will breach the capital adequacy requirements.

    In this context, one must consider also that Directives Nos 2009/111/EC, 2010/76/EU, and Regulation (EU) no. 1205/2011, which seem aimed to improve the consolidation of securitization vehicles and the disclosure requirements, in order to improve financial market transparency.
53. See Vento and La Ganga (2009) "Bank Liquidity Risk Management and Supervision: which Lessons from Recent Market Turmoil?," *Journal of Money, Investment and Banking*, Issue 10, p. 78 ff.
54. I refer to the role of the "Euro Summit" (art. 12, Tr. SCG) and of the "Conference of the Representative" (art. 13, Tr. SCG); taking into account also the effect of the well-known *"balanced budget rule"* (art. 8, Tr. SCG).
55. See Pellegrini (2003) *Banca Centrale Nazionale e Unione Monetaria Europea*, (Bari), p 235 ff.
56. See Barbagallo (2014) "L'Unione Bancaria Europea," speech at NIFA—New International Finance Association, p. 3, where the authors assess the reason of this choice.
57. Underlying the above there is an idea of the unsustainable asymmetry between money and credit, see Masera (2013) *Moneta europea credito nazionale*, in *La Repubblica*, June 17, 2013. See also De Polis (2014) "Unione Bancaria e gestione delle crisi: un modello di Banca in trasformazione," *XII PAN European Banking Meeting*, p. 5 ff., for the pillars of banking crisis management.
58. See Ferrarini and Chiarella (2013) "Common Banking Supervision in the Eurozone: Strengths and Weaknesses," *ECGI—Law Working Paper*,

no. 223/2013, cit.; see also Austrin-Willis (2011) "EU and U.S. Solutions to Systemic Risk and Their Potential Influence on a World Trade Organization Approach," *Georgetown Law and Economics Research Paper*, where the author analyzes these two approaches to addressing systemic risk, which—in his opinion—differ greatly in both their specificity and the level of authority they entrust to centralized regulators. It is interesting that the idea of an "EU approach is better suited for adaptation to the WTO."

59. See Schoenmaker (2014) "The New European Banking Union Landscape," *Duisenberg School of Finance Policy Brief*, no. 35, where the author explains how the advance to Banking Union may creating a truly integrated EUM internal market.
60. See Hellwig (2014) "Yes Virginia, There is a European Banking Union! But it May Not Make Your Wishes Come True," *Max Planck Institute Collective Goods Preprint*, no. 2014/12 for a critical review of the prospects for European Banking Union as they appear in the summer of 2014.
61. See Henry et al. (2013) "A Macro Stress Testing Framework for Assessing Systemic Risks in the Banking Sector," *ECB Occasional Paper*, no. 152., where it is explained why the use of macro stress tests to assess bank solvency has developed rapidly over the past few years.
62. See European Commission, Press release, Brussels, January 23, 2014, p. 2.
63. See, on this point, Cassese, "Produzione delle regole e allocazione dei poteri di vigilanza nella dimensione europea," speech at the Congress "Dal Testo unico bancario all'Unione bancaria: tecniche normative e allocazione dei poteri," Rome, September 16, 2013.
64. It is well-known that ECB went beyond the euro system's regular open market operations, which consist of one-week liquidity-providing operations in euro (main refinancing operations, or MROs) as well as three-month liquidity-providing operations in euro (longer-term refinancing operations, or LTROs). This, with LTROs, provides additional, longer-term refinancing to the financial sector.

    It is important, for these purposes, to consider that in September 2014 the ECB announced two new purchase programmes, namely the ABS purchase programme (ABSPP) and the third covered bond purchase programme (CBPP3). The programmes will enhance transmission of monetary policy, support provision of credit to the euro area economy and, as a result, provide further monetary policy accommodation; see ECB, *Open market operations*, available at http://www.ecb.europa.eu.

    This is in line with the goal of Draghi (2014) *Monetary Policy in a Prolonged Period of Low Inflation*, Sintra, May 26, 2014, where he clarified that "an intermediate situation is one where credit supply constraints interfere with the transmission of monetary policy and impair the effects of our intended monetary stance. This would require targeted measures to help alleviate credit constraints."
65. See, on thispoint, Capriglione and Semeraro (2012) *Crisi finanziaria e dei debiti sovrani. L'Unione Europea tra rischi ed opportunità* (Torino) p. 100 ff., where the analysis is recalled of Belke (2012) "Three years LTROs—A first assessment of Non-Standard Policy," *Note of the Directorate General for Internal Policies—Policy Department: Economic and Scientific Policy*, April.

66. See Luck and Schempp (2014) "Banks, Shadow Banking, and Fragility," cit., where it is clarify that a safety net for banks may fail to prevent a banking crisis, but it may become too costly for the tax payers.
67. See FSB, *Recommendations to Strengthen Oversight and Regulation of Shadow Banking*, October 27, 2011, p. 16 ff.
68. See Zetzsche (2012) *The Alternative Investment Fund Managers Directive – European Regulation of Alternative Investment Funds* (Alphen and den Rijn), p. 21 ff., where the author highlight that "when the financial crisis came upon Europe…the light touch regulation, pro-market attitude for which the hedge funds group and the private equity group stood ran counter to the political interests of subjecting managers of AIFs to more stringent regulatory oversight."

    For an interesting point of view, see also Buttigieg (2013) "Negotiating and Implementing the AIFMD: The Malta Experience," *The Accountant, Spring 2014*, p. 34 ff.

    With regard to UCITs, see European Commission, *Greater Protection for Retail Investors: Commission Welcomes European Parliament Adoption of Strengthened European Rules on UCITS*, Brussels, April 15, 2014.
69. See Directive 2011/65/UE, Recital no. 94
70. See Regulation (EC) no. 2533/98; Regulation (EC) no. 24/2009; Guideline ECB/2008/31.
71. See Coulter (2013), "Capital Recycling and Moral Hazard in the Securitization Market," *2013 Financial Markets & Corporate Governance Conference*, where the author considers if rating agencies may mitigate the moral hazard problem of banks and propose a model to understand the interaction between a bank, a credit rating agency, and a continuum of investors, considering that the presence of a credit rating agency may still improve social welfare.

    On the contrary, the regulators seems oriented to further requirements to promote the maximization of social welfare, based on the retention models; see on this point the interesting analysis of Malekan and Dionne (2012) "Securitization and Optimal Retention Under Moral Hazard," *SSRN Working Paper*, no. 2038831, where the authors show that the optimal contract must contain a retention clause in the presence of moral hazard.
72. On the prospective for a cognitive-based approach to the regulatory process, see Di Porto and Rangone (2013) "Cognitive-based regulation: new challenges for regulators?—Regolamentazione su base cognitiva: nuove sfide per i regolatori?," *federalismi.it*, f. 20, p. 29 ff.; Rangone (2012) "Errori cognitivi e scelte di regolazione," *Analisi Giuridica dell'Economia*, f. 1, p. 7 ff.; Rangone (2012) "Il contributo delle scienze cognitive alla qualità delle regole—Behavioral science findings to improve the quality of regulation," *Mercato concorrenza regole*, f. 1, p. 151 ff.

    See also Siclari (2007) "Gold plating e nuovi principi di vigilanza regolamentare sui mercati finanziari," *Amministrazione in cammino*, where the author focus on the need for a level playing field in the financial markets.
73. See European Commission, "Proposal for a Regulation of the European Parliament and of the Council on 'Reporting and transparency of securities financing transactions'," Brussels, January 29, 2014 COM(2014) 40 final2014/0017 (COD) and, in particular, its explanatory memorandum,

where it is highlighted that "the crisis highlighted the need to improve transparency and monitoring not only in the traditional banking sector but also in areas where non-bank credit activities took place, called shadow banking."

74. See, on this point, Cassese (2009) *Il diritto globale*, (Torino) p. 17 ff.
75. See Lemma (2006) " 'Soft law' e regolazione finanziaria," *Nuova giuris prudenza civil ecommentata*, f. 11, p. 600 ff.; see also Gersen and Posner (2008) *Soft Law, U. of Chicago, Public Law and Legal Theory Working Paper*, no. 213, where the authors clarify that soft law consist of "rules" issued by bodies that do not comply with procedural formalities necessary to give the same an "hard legal status," but nonetheless these rules may influence the behavior of other law-making bodies and of the public.
76. See Claessens, Pozsar, Ratnovski, and Singh (2012) *Shadow Banking: Economics and Policy*, in IMF Staff Discussion Note.
77. See Ghosh, Gonzalez del Mazo, and Ötker-Robe (2012) "Chasing the Shadows: How Significant is Shadow Banking in Emerging Markets?", *The World Bank Poverty Reduction and Economic Management Network—Economic Premise*, where the authors specify that in select Central Eastern European countries, shadow banking grew rapidly until 2007, and then lost some of its share following the global financial crisis.
78. It is helpful to recall an analysis on the relations between the International financial centers and the viability of certain responses to the financial crisis of 2007–2009, and to the market global economic deceleration that ensued, agreed in 2009 by governments and financial authorities of the more developed countries. This, with particular regard to the ones that take into account the common strategies to overcome the crisis; see Santillan-Salgado (2011) "Offshore Financial Centers: Recent Evolution and Likely Future Trends," *Journal of Global Economy*, Vol. 7, Issue 2, where the author highlight the danger in getting to adopt measures that may dampen economic recovery by discouraging the flow of credit to the real economy.
79. See World Bank, *What the World Bank Is Doing*, available at http://www.worldbank.org, where it is highlighted that, in addition to its traditional tasks, the World Bank is also helping countries build resilience to external shocks associated with market volatility, by facilitating access to market-based risk management tools and capital market solutions.
80. One must take into account the position of Tobias (2011) "Monitoring Risk in the Shadow Banking System," *11th Annual International Seminar on Policy Challenges for the Financial Sector*, where it is assumed that much of the external shadow banking system can be supervised indirectly via banks or dealers, and then there is the conclusion that macro-prudential supervision includes the interaction of supervised institutions with shadow banking activities.
81. See also Sharma (2014) "Shadow Banking, Chinese Style," *Economic Affairs*, Vol. 34, Issue 3, p. 340 ff., where he tries to measure and understand the growth of shadow banks in China, by investigating how Chinese Government can best utilise the services of shadow banks without create systemic risks for the global financial system.
82. On the opacities of the "China's Shadow Banking," see Zhanh (2014) *Inside China's Shadow Banking. The Next Subprime Crisis?* (HI—USA), p. 95 ff.

83. See Ghosh, Gonzalez del Mazo, and Ötker-Robe (2012) "Chasing the Shadows: How Significant is Shadow Banking in Emerging Markets?," cit., p. 6.
84. See IFM, *How We Do it. Economic and Financial Surveillance*, available at http://www.ifm.org; see also G20, *Reforming Global Institutions*, Australia 2014, available at http://www.g20.org, where it is highlighted that "reforming the IMF will ensure that it adapts to changing environments and remains strong, influential and representative of the global economy."
85. See Errico et al. (2014) "Mapping the Shadow Banking System Through a Global Flow of Funds Analysis," *IMF Working Paper—Statistics Department*, p. 37, even if it is still to be completed the data collection, in order to proceed, successively, to the verification of their reliability, as well as to identify the nodes interconnected of financial flows mentioned above (with respect to "each sector in each location").
86. One should also consider that the regulatory efforts need to focus on the sizable volumes of bank funding coming from non-bank asset managers via source collateral and institutional cash pools; see Pozsar and Singh (2011) "The Non-Bank-Bank Nexus and the Shadow Banking System," *IMF Working Paper—Research Department*, p. 14.
87. See IMF (2014) *Global Financial Stability Report. Risk Taking, Liquidity, and Shadow Banking: Curbing Excess While Promoting Growth*, cit., p. 74, where the Fund identifies the key drivers of the growth patterns of the shadow banking system and quotes the results of Jackson (2013) "Shadow Banking and New Lending Channels—Past and Future," *50 Years of Money and Finance: Lessons and Challenges*, Vienna: The European Money and Finance Forum; Caballero (2010) "The 'Other' Imbalance and the Financial Crisis," *NBER Working Paper*, no. 15636; Goda and Lysandrou-Stewart (2013) "The Contribution of U.S. Bond Demand to the U.S. Bond Yield Conundrum of 2004 to 2007: An Empirical Investigation," *Journal of International Financial Markets, Institutions and Money*, Vol. 27, pp. 113–136; Goda and Lysandrou (2014) "The Contribution of Wealth Concentration to the Subprime Crisis: A Quantitative Estimation.," *Cambridge Journal of Economics*, Vol. 38, Issue 2, pp. 301–327; and Lysandrou (2012) "The Primacy of Hedge Funds in the Subprime Crisis," *Journal of Post Keynesian Economics*, Vol. 34, Issue 2, pp. 225–253.
88. See Skeel (2010) "The New Financial Deal: Understanding the Dodd-Frank Act and its (Unintended) Consequences," cit., where the author highlights the implications for derivatives of the US regulation, will impose a new order (on this previously unregulated industry) and, in particular, it require that most derivatives shall be traded on an regulated trading venues (and backstopped by a clearing house or a central counterpart).
89. See Marchesi and Sabani (2013) "Does it Take Two to Tango? Improving Cooperation between the IMF and the World Bank: Theory and Empirical Evidence," *Centro Studi Luca d'Agliano Development Studies Working Paper*, no. 357, for an empirical analysis that shows that a Bank–Fund simultaneous intervention is beneficial to growth and that such beneficial effect is increasing with the willingness to coordinate of the two organizations. According to these Authors, this evidence would be in favor of a (more) centralized governance.

90. See Fitoussi and Laurent (2008) *La nuova ecologia politica* (Milano) p. 9, where the authors describe the economic self-regulation as the failure of an illusion.
91. See FSB, *2012 Global Shadow Banking Monitoring Report*, cit., p. 6; see also FSB, *2014 Global Shadow Banking Monitoring Report*, cit., p. 24, which confirms that systemic risks can spill over from shadow banking entities to the banking sector.
92. See FSB (2011) "Recommendations to Strengthen Oversight and Regulation of Shadow Banking," cit., p. 8.
93. The width of the "scope" therein corresponds to the option for a "process" that can be summed up in a monitoring framework for the shadow banking system able to identify and assess the risks on a regular and continuous basis; follows the input according to which the relevant authorities should have powers to collect all necessary data and information, as well as the ability to define the regulatory perimeter for reporting as well as the need that authorities should ex-change appropriate information both within and across the relevant jurisdictions on a regular basis to be able to assess the risks posed by the shadow banking system.

    In general, I shall take into account the opinion of Sepe (2014) "A Crisis, Public Policies, Banking Governance, Expectations & Rule Reform: When Will the Horse go Back to Drink?," *Law and Economics Yearly Review*, Vol. 3, Part 1, p. 210 ff., on the limits of current public policies.

    According to the above opinion, clearly there is a preference for a model of supervision based on the activities of scanning and mapping, in which the identification of the aspects of the shadow banking system posing systemic risk or regulatory arbitrage concerns will require a specific combination of intervention policies.

    In the end, I agree with the decision of FSB proceeding to the regulation of banks' interactions with shadow banking entities; see FSB (2011) "Recommendations to Strengthen Oversight and Regulation of Shadow Banking," cit., pp. 4–5 and 7–12.
94. See Dallas (2011) "Short-Termism, the Financial Crisis, and Corporate Governance," *Journal of Corporation Law*, Vol. 37, p. 264 ff., for an exploration of the reasons why financial and non-financial firms engage in short-termism with particular attention given to the financial crisis of 2007–2009.
95. See Posner (2007) *Economic Analysis of Law*, cit., p. 469, where the author explains the link between diversification, leverage, and debt-equity ratio and p. 473 on the challenge of "behavioral finance," which—as aforementioned—does not find any protection in the shadow banking system.

    It is important to recall the author's idea that "it is important that the law not assume that people are more rational than they are" (p. 475).
96. This is consistent with the approach of Shavell (2004) *Foundations of Economic Analysis of Law* (Cambridge), p. 473 ff., and, in particular, pp. 474–490, where there are the differences between the "certain enforcement" and the "enforcement with a probability."
97. See Draghi (2014) "Financial Integration and Banking Union," speech at the Conference for the 20th Anniversary of the Establishment of the European Monetary Institute, Brussels, February 12, 2014, where he discussed the

question whether the quality and comprehensiveness of integration matters. And, in his opinion, they do. Draghi concludes that there are costs which can arise from a type of financial integration (that is short term and reversible); or from having perfect integration in one market and fragmentation in another.
98. See Cooter and Ulen (200) *Law and Economics*, (Boston) p. 91, where there is an analysis of the key elements of transaction costs, and p. 95 ff. where the authors focus on the level of them and the appropriate legal rule.
99. See FSB (2011) "Recommendations to Strengthen Oversight and Regulation of Shadow Banking," cit., p. 12, and FSB, *2014 Global Shadow Banking Monitoring Report*, cit., where the overview of the global macro-mapping results shows with emphasis that the "Monitoring Universe of Non-Bank Financial Intermediation" continued to grow in 2013.
100. See FSB (2014) *Structural Banking Reforms: Cross-border Consistencies and Global Financial Stability Implications*, for a quantitative assessment of potential cross-border financial stability implications related to national structural banking reforms that have recently been implemented or proposed.

Index

adverse selection, 43
AIFMD, 24, 63, 73, 106, 173
algorithmic trading, 29, 155
Alternative Investment Funds, 73, 173
anti-avoidance rule, 35
anti-terrorism, 32
areas of risk, 130–133
asset manager, 39, 73–6, 93, 106,
Asset Quality Review, 66, 158
asset-backed commercial paper, 2, 18, 68, 117–120
asset-backed securities, 2, 22, 68, 117–120

backstops, 27, 51, 134, 141
Bail-in, 156
balance sheets, 44–46, 128, 172
Bank of England, 18, 21, 76, 104, 115, 152
Bank of Italy, 18, 21, 66, 69
bank, 9, 21, 23, 64–65, 78, 81–7, 127–129, 148
banking services, 81, 102, 127, 169
Basel, Accord of, 4, 31, 35–6, 51, 65, 82, 101, 104, 129, 133, 159, 177, 180
black box, 68
black market, 2, 31, 34
black-ops, 125
boundaries, 12, 27–28, 31–6, 98
BRRD, 156
Bundesbank, German, 21, 157

capital adequacy, 4, 22, 36, 76, 129, 131–2, 184
Capital Markets Union, 125, 161–2, 183
captive financial company, 63, 71
cash pooling, 40, 90, 113
cash, 32–33
Central banks, 21, 31, 52, 151–2, *see also under individual names*
China, 25, 176

collateralized borrowing, 113–4, 125
collateralized debt obligations, 2, 9–10, 45, 68, 117–120
competition, 19, 26, 35, 40–41, 55, 84, 89, 98, 111, 118–9, 126, 151, 159, 161, 182
compliance function, 86, 91, 94
conduits, 16, 71, 97
conglomerates, 82
contamination risk, 47, 134–6, 147
contracts, 2, 20, 42, 54, 83, 96, 99–101
corporate governance, 53, 65–7, 86, 142–5,
counterparty risk, 54, 84, 123, 129, 133, 145
CRD IV, 25, 35, 129, 159
credit enhancement, 71, 95, 98, 118, 132, 143, 146
credit institutions *see* banks
Credit Rating Agencies, 108–112
credit transformation, 101, 107, 120, 133
crisis, financial, 5, 11, 147
cross-board execution, 52
CRR, 66, 144–5, 159

De Larosière report, 160
Definition of the shadow banking system, 1, 13–22, 25, 26, 30, 38, 48, 74,
deregulation, 47, 154–7, 182
derivatives, 120–124, 150 *see also under individual names*
Discount window facility, 152
Dodd-Frank Act, 177

EBU *see* European Banking Union
EBU *see* European Banking Union
EC *see* European Commission
ECB *see* European Central Bank
ECB *see* European Central Bank
economic determinants, 4, 37–44, 156–158

efficiency, 12, 30, 41, 44, 57, 71–73, 77
EFSF *see* European Financial Supervision System
eligible counterparties, 74
emerging countries, 26, 162, 176
EMIR, 19, 123, 129, 162
EMU *see* European Monetary Union
entities, 62–77, 70–73, 133
equilibrium, 63, 85, 131, 139
ESAs, 164, 165–167
ESCB, 173
ESM *see* European stability Mechanism
ESRB, 163–4, 170
EU *see* European Union
Euro Summit, 161
European Banking Union, 14–15, 86, 154, 167–171, 183
European Central Bank, 21, 67, 74, 80, 104, 152, 154, 157, 167
European Commission, 23, 48–49, 75, 142, 161–163
European institutions, 23, 160–161 *see also under individual names*
European internal market, 158–160
European Monetary Union, 21, 35, 124, 167
European stability Mechanism, 125, 160
European System of Financial Supervision, 25, 125, 139, 154, 163–164, 166, 168
European Union, 20, 125, 158, 160–3, 168, 183
exogenous risks, 147–151
expenses, 130

fair value, 46, 94, 108
FED *see* Federal reserve system
Federal Deposit Insurance, 27
Federal reserve system, 21, 26, 30, 75, 80, 88, 99, 152
financial innovation, 56, 171, 182
financial instrument, 39, 117–124, 149, *see also under individual names*
Financial Stability Board, 6, 17–19, 30, 53, 65, 146, 178–181
financial supervision, 8, 41, 74
financial transactions, 13
financialization, 11, 38

Fiscal Compact *see* SCG Treaty
forwards, 121
freedom, 53, 58–61, 139
FSB *see* Financial Stability Board
fundamental rule, 13
funding, 23, 38, 148
futures, 3, 121, 150

G20, 5–7, 20, 56, 104, 177
global financial system, 11, 54, 97
global regulators, 178, 180, 184
globalization, 10, 55–58, 157
governance risk, 142–145
government oversight, 120, 155
guarantees, 84, 88

high-frequency trading, 29, 155

IMF *see* International Monetary Fund
incentives, 39, 56, 107
Information asymmetries, 42–44, 45, 139
information needs, 43, 141
instability, 41, 44–48, 60, 130, 159, 164
insurance company, 101
interlocking directorates, 45
intermediation margin, 4, 40
internal audit, 86, 91
internal controls, 59, 78, 85–87, 94, 143
International Monetary Fund, 13, 20, 40, 175–178
internationalization, 52
ISDA, 123
ISLA, 114–115
issuing, 62, 106, 113, 119–120

leveraging, 47, 107, 143, 148, 175,
liberalization, 16, 59, 73, 126, 156
liquidity, 27–28, 37, 105, 140
loan, 45, 50–1, 63, 70, 83–4, 117, 151

macro-prudential supervision, 24, 163–164, 171
mapping, 19, 177, 179
mark-to-market, 46
maturity, 40, 101, 106, 115
MiFID II, 14, 24, 56, 162, 172–175

MiFID, 24, 56, 155
MiFIR, 25
monetary authorities, 8, 67, 113, 148
monetary policies, 98, 151–153, 172
money laundering, 13, 31, 57
money market funds, 20, 75–77
monitoring, 41, 48, 67, 171, 179
moral hazard, 60, 135, 146, 173
movement of capital, 1, 14, 136, 154
multi-seller, 71

non-standard operations, 113–129
NSOs *see* non-standard operations

offering, 117–120
opacity, 44–46, 100, 151
operators, 9, 42, 50, 52, 79, 136, *see also under individual names*
opportunistic behavior, 31, 49, 70, 135
options, 3, 102, 121, 150
organizational risks, 139
organizational structure *see* corporate governance
originate-to-distribute, 10, 50, 82, 105, 132
OTD *see* originate-to-distribute
own funds, 90, 35, 166

pension funds, 78, 92–94
performance, 41, 44, 132
pooling of loans, 45
premium, 88, 128
pro-cyclicality, 44–47, 116
prudence, 76, 86, 136
public authorities, 28, 46, 155, *see also under individual names*
public intervention, 30, 53, 146, 151, 154, 183

rating, 44, 83, 107–112
regulators, 12, 26, 175, *see also under individual names*
regulatory capture, 49
repo-market, 47
repo, 68, 113–117

Reserve Bank of India, 25
resilience, 133, 163, 179, 182
Return on Equity, 51
risk factors, 132–136, 140
risk management, 94, 105, 136–139
risk, 4, 48, 55, 83, 89, 91, 112, 130–153, 179
roll over risk, 135
runs, 141, 148

safeguards, 85, 140–142
savior of the last resort, 126
SCG Treaty, 125, 160
scoring, 83
securitization, 3, 63, 83, 90, 102–105, 149
security, lending and borrowing, 113–117
Seoul Action Plan, 5
shadow bank, 27, 67–70, 96
shadow banking sub-system, 29, 50, 82
shadow business, 78–94, 82
shadow credit intermediation process, 2–3, 9, 62, 95–98, 114, 120
shadow dynamics, 28
shadow funds, 73–74
shadow operations, 2, 35, 95–112
shadow preferences, 37
shadow, 3, 15
SIFIs *see* systemically important financial institutions
Single Rulebook, 35, 154, 173
Single Supervisory Mechanism, 167, 169
single-seller, 71
soft law, 49, 113, 165
solvency II, 88, 90–91, 104
sovereign debt, 124–127
sovereign derivatives, 124
special purpose vehicle, 2, 45, 62–67, 102, 127
SPV *see* special purpose vehicle
stability, 5, 23, 40, 156, 166
stock exchange, 55
stress tests, 170
structured investment vehicles, 16, 72
supervision, 30, 138, 141, 154–181

supervisory system, 132, 156–158
sustainability, 10, 49, 139
swaps, 3, 121, 124
synthetic transactions, 59
systemic risk, 18, 122, 137, 174, 179
systemically important financial institutions, 6, 54, 146, 159

tax evasion, 13, 31–33
technical reserves, 88
TLTROs, 151, 157, 171–172
too big to fail, 145

transformation, 42, 64, 101–107, 118, 121, 131
transparency, 43, 61, 108, 138, 149, 166, 175, 181

UCITS, 73–75, 106, 111
uncertainty, 47, 131–132, 161
United States, 52, 75, 181
US SEC, 75

vulnerabilities, 21, 130

wholesale funding, 52, 97–98
World Bank, 175–178

CPSIA information can be obtained
at www.ICGtesting.com
Printed in the USA
LVOW13s1447100317
526809LV00020B/133/P